William Wirt Kinsley

Science and prayer

William Wirt Kinsley

Science and prayer

ISBN/EAN: 9783337283216

Printed in Europe, USA, Canada, Australia, Japan

Cover: Foto ©berggeist007 / pixelio.de

More available books at **www.hansebooks.com**

Chautauqua Reading Circle Literature

SCIENCE AND PRAYER

BY

W. W. KINSLEY

Author of "Vexed on Vexed Questions."

FLOOD AND VINCENT
The Chautauqua-Century Press
MEADVILLE PENNA
150 FIFTH AVE. NEW YORK
1893

CONTENTS.

Chapter.		Page.
I.	God can Interfere when and as He Chooses, without Destroying any Force or Abrogating any Law	5
II.	He has in fact from time to time thus Interfered and is still Interfering	27
III.	He will Interfere for each one of us, however Insignificant we may at present seem to be	44
IV.	He will Interfere because we ask Him, Doing for us what otherwise he would not have Done	67
V.	Every Sensible Prayer offered in the Right Spirit is certain of Favorable Answer	94

SCIENCE AND PRAYER.

I.

THE Scriptures affirm, that, in answer to prayer, a part of Palestine was once visited with long drought and afterward with copious rains and harvests, an entire family healed, a raging fire quenched, God's purpose to destroy a stiff-necked people changed, the sun and moon apparently stopped in mid-heaven for an entire day, a thunderstorm made to burst right in wheat harvest, a leprous hand cured, a dead child revived, a good king's life lengthened, and, for an assuring token, a dial's shadow actually turned backward.

The Bible unmistakably teaches that God both can and does interfere in our behalf, that his interference often is a direct result of our asking, that all reasonable prayers offered in a right spirit are certain of favorable answer. The requests may be as varied as the healthful and intelligent longings of human hearts.

Some scientists smile at what they style the childish credulity of the Christian's creed. Our investigations, say they, have disclosed a universal reign of unchangeable law, not only in the production of material but even of mental phenomena. We have found that within the walls of every particle of matter there is lodged a force; that these forces are of sixty-four or more different kinds, and their differences in nature and effect make all the differences in the substances about us; that they bear to each other certain unalterably fixed relations, and exert over each other unalterably fixed influences. These relations we have been able by our experiments to reduce to mathematical formulæ. We have found that these forces never manifest themselves unless certain conditions are fulfilled, and that, when they are, the forces invariably appear and act always in precisely the same way. It is also claimed, that, as far back as we can peer into the past, this same order has prevailed; that this rock-ribbed, wave-washed, verdure-clad, densely populated earth of ours has come up out of chaotic fire-mist by the operations of none other than these very forces which at the first were hidden within it; that the earth has developed from its unorganized primordial state into its present complexity with as regular gradations of growth as those through which the oak passes in pushing up from out the walls of the acorn its sinewy stem with outreaching boughs and waving pennons; that the earth itself is an organism as truly as the tree, has like complemental parts, has had a germinal beginning, has been, and still is, incarnating under pre-established laws of evolution, point by point, age after age, a certain set ideal under the guidance of a central germ-power, divinely commissioned it may be, but commissioned even as to the details of its finest microscopic work, untold millions of years ago.

How idle, then, it is, they claim, for weak, blind children of a day to presume to break in on this grand order of the universe! Go out into nature, they tell us, and you will find that not a single one of her laws is ever abrogated, that from their control not the least thing is for an instant released. Gravity holds in its grasp not only the ponderous suns with their whirling satellites, but every infinitesimal mote that floats in the air. The force shut up within the walls of an atom of carbon is never dislodged, and never loses a single characteristic. Manacle it with fetters of frost, immerse it in the white heat of a furnace, smite it with a trip-hammer on the face of an anvil, hurl it into the chemical embrace of an affinitive element, do what you will with it, it will reappear identically the same atom informed by precisely the same mysterious force. This speck of matter defies all powers of earth or sky to batter in its walls and drive out its occupant. Every force, the world over, says that only those who find its secret and meet the conditions can command its services. Do you want bread? Here are the seed, the soil, the air, the shower, and the sunbeam. Matter and force are at your bidding, but their laws are inexorable. Rays of light will travel ninety-five millions of miles to serve you; the atmosphere will gather its clouds from the ocean and float them across a continent to pour their treasures at your feet; the mountains will furnish you millstones, and the running brooks will turn them. The forests that grew a hundred thousand years ago you may find packed away in beds of anthracite, waiting to heat your ovens so soon as your dough is ready for the baking. Not a force in nature but will serve the veriest outcast if he will comply with the conditions; not one, even the humblest, will condescend to move so much as a hair's-breadth even for the Czar of all the Russias, unless he does. The prayerless sinner and the praying saint meet here on a common level. All those stories about producing thunderstorms by prayer, healing the sick, turning back shadows, stopping the sun in the heavens, raising the dead, are thoroughly unscientific and absurd, and the height of absurdity is reached when it is claimed that the all-wise Creator can be induced to change his plans by the importunate pleadings of a little creature to whom he has given a brief existence on one of the obscure satellites of one of the million suns that make up one of the nebulous clusters with which the heavens are swarming. What greater presumption can be imagined? Has the Almighty so sadly blundered in his plans that this little creature can discover to him their defects, and induce him to make a change at this late day, when everything is so intimately interlinked and interdependent that an interference in one part may demand a reconstruction throughout the whole in order to avoid widespread confusion and ruin? Can God spare any special thought now for such infinitesimal interests so long as the concerns of this vast swinging universe are upon him? He has laid down broad general plans. We cannot reasonably expect him to listen to our baby prattle about the petty details of our vanishing lives. If we thrust our hands into the fire, live in a malarious district, are capsized in mid-ocean, we must suffer the natural consequences, and look about us, as best we can, for a more congenial environment.

Such, in brief, is the attitude assumed at the present day by a majority of scientists on this one of the most vital and perplexing of questions. This their creed is, as I think can be clearly shown, a most mischievous mixture of truth and error. The spirit of cold speculative scepticism pervading it is making rapid inroads upon all classes in society. How many even of those who have been gathered into the fold of the church have fallen under the blighting spell of this genius of modern materialistic thought! How many prayers are simply the outbreathings of a reverential fear, or are a mere dead formalism, or the results of sheer habit! How many are little else than agonized longings accompanied with no joyous expectation! How few, very few, are offered with the same confident assurance of results as inspires the farmer when he sows his fields, or the telegraphic operator when with his key he closes the electric circuit and sends his messages over the long leagues of ocean cable !

My purpose at present is to show :—

1st. That phenomena and the producing forces with their laws or modes of working, brought to light by scientific investigations in the fields of physics and of metaphysics, harmonize perfectly with the Scripture view of prayer, and abound in suggestions of how God can interfere in nature without destroying any force or abrogating a single law.

2d. That, as a fact, he has thus actually interfered again and again.

3d. That it is not only not presumptuous, but most natural and reasonable, for us to expect that he will interfere for *us*, insignificant though we may seem to be.

4th. That he will interfere because we ask him, doing for us what otherwise he would not have done.

5th. And, lastly, that he will not in a single instance withhold any real blessing which is asked for in the right spirit, and the bestowal of which lies within the compass of his power.

1st. *How* can God answer prayer without destroying any force or abrogating any law? In my own experience, real light on this point first came from the perusal of Dr. Bushnell's "Nature and the Supernatural." His mode of treatment has long since passed out of memory, but a thought-germ was lodged in my mind which has since grown into a deep-rooted conviction. As, however, I have followed out these lines of thought, it has been a constant source of surprise that so many of the scientists, while they have with tireless patience and keenest insight unraveled much of the infinite intricacy that attends the interplay of nature's forces, unearthing so many secrets and becoming masters in so many fields of inquiry, have seemingly lost sight of that most interesting and important of all facts, that everywhere ample provision has been made for the efficient interference of direct will power. They of course cannot have failed to discover it, for there is hardly a waking moment in the lives of any of us when we are not conscious that we actually exercise volitions, and that these volitions effect changes, and sometimes most important ones, in the world about us. How our wills are thus linked with matter, it would probably puzzle the wisest to explain; but that they actually are, is a fact patent to all. And so I surmise it is not the fact, but the deep significance of the fact, that has so strangely escaped the notice of so many of our savants of science.

Over my body in many particulars my will exercises direct control. I, for instance, order my hand lifted. The mandate instantly flashes from the brain down the motor nerves to the very muscles in waiting, and their fibres at once begin to shorten. I exercise this direct will power right against the force of gravity, temporarily overpowering but not destroying it, for it still continues to pull the hand down with the same might as before. This overbalancing of one force by another is taking place everywhere throughout nature. For illustration, take a tumbler of water. If it were not for the cohesive attraction between the particles of the glass being stronger than the gravity, the sides would crumble into dust, and sink with the water to the lowest attainable level. Gravity has not been destroyed, but simply mastered by a stronger antagonist. Remove a part of the heat from the water, and it will become a crystallized solid, showing that until now the heat force has been holding the crystalline in check. Lower still further the temperature, and the sides of the tumbler will burst in pieces, the crystalline force overcoming the cohesive. Raise the temperature, and the water will change to steam, and a repellence between the particles will appear, the heat driving them asunder, despite all that cohesion and gravitation can do.

Over the world outside the body, the control of our wills, though mostly indirect, is equally potent, and yet nature is not thrown into confusion, not a single force destroyed, not a law abrogated. Our volitions are simply supernatural, not contranatural. Our wills act indirectly by complying with the conditions that unfetter nature's forces. The scientists have established beyond question the fact that there is not a single one of these forces that is not wholly inoperative unless certain conditions are fulfilled, and just as soon as they are, the force begins to work its wonders. Scientists have even gone further, discovering in very many instances precisely what those conditions are, and thus placed it within our reach to utilize those forces in the arts of life.

Back of our will power, acting as its guide, there now exists, thanks to these explorers, a well-informed intelligence, and we have become masters of nature by simply understanding and complying with her laws. For instance, we want homes for ourselves and our little ones, and so we cast about and find abundance of crude material,—sand and clay, metal and slate, rock and standing trees and running water. Our wills decree that these shall be transformed into cemented walls of brick and stone, framed timbers, tessellated floors, frescoed ceilings, plate-glass windows, roofs and mantels, furnaces and swinging doors, and step by step, under the quickening power of the mind, the wondrous change is wrought. We even make our wills felt in the domains of vegetable and animal life, improving old varieties and developing new ones among fruits and flowers and domesticated animals, enriching and seeding our soils, and multiplying our flocks and herds to meet our ever-growing wants.

The processes by which our wills enforce their decrees may be a little tedious, but the ends are reached, the course of nature is seriously broken in upon, results attained which otherwise nature never would have attempted, yet no disorder has anywhere ensued. What marvelous effects have been produced by this intelligent will power of man, cunningly directing to its own uses the ever-waiting elemental and vital forces! How many rivers have been bridged, beds of rivers shifted or tunneled, mountains discrowned or their rocky centers pierced to open highways for the world's commerce! The very lightnings have been tamed into flying Mercurys to carry the thought-messages of this busy-brained master, the oceans whitened with his sail, the continents covered with his networks of railways and canals, barren wastes changed into vineyards and palm-groves and orange-orchards, the unshapely quarries of granite and of marble transformed into palaces and statue-crowned temples to body forth his ripest culture and most holy thought.

The influence of the human will has had even a wider circuit assigned it. Many of us have known instances of weak wills being overawed by stronger ones, and the domination being so absolute as for the time being to actually blot out every distinctive trace of personality and suspend individual responsibility. Not one of us but has felt, time and again, the indirect power of another's will reaching us through channels of argument, persuasive kindling of the fancy, eloquent appeal, shrewd suggestion, or show of appreciative sympathy.

There are a thousand avenues to the heart, a thousand ways to arouse the conscience, inflame passion, fill the chambers of the soul with dread alarms, and these are discovered and utilized by positive and aggressive souls athirst for wealth, power, or prestige. Society has its born leaders. Individuality and responsible free choice are with the vast majority still retained, but it is through these multiform influences of personal character that the life of the world's subtile social organism is, under pre-established spiritual laws, regulated and maintained.

Thus we see that to the touch of the human will all nature is plastic, that every facility has seemingly been provided for its efficient interference. Think you that, in a world where so many doors have been so invitingly left open for the will of the creature to enter and occupy, the will of the Creator has been studiously excluded? Can science, which has so conclusively proved the one, consistently deny the other? Is it not rather forced to assert that, so far as God's will has greater innate power and is guided by a profounder knowledge, it has proportionately greater facilities for effecting its purposes and, at the same time, leaving every force and law both in the material and mental kingdoms equally undisturbed?

Before the birth of science a radical misconception of the true nature of miracles was entertained, and seems still very generally to prevail, and this has doubtless largely provoked the attacks made on the truthfulness of the Bible record. Cannot the miracle-workings spoken of have been wrought by acts of divine will precisely analogous to those of the human? What necessity is there for thinking that any force or law has been, or need be, destroyed? The ax that was made to float on the water by God's command through his prophet was not necessarily made lighter than the water any more than my hand when I raise it is made lighter than the air. The nature of the materials remained the same, and gravity was still in full force, but God's will was under the axe as mine is under the hand. Precisely how it got there I cannot explain, neither can I how mine got under the hand. The one is no more mysterious than the other, no more of a deviation from nature's laws, but both volitions are, as far as I can discover, essentially the same.

There perhaps is no Bible narrative whose truth has been more violently and generally assailed than that of the sun's being stayed upon Gibeon and the moon in the valley of Aijalon. It has been pronounced scientifically impossible. Some authors have attempted to explain it by claiming that it is only a quotation from the Book of Jasher, a mere poetical extravaganza embellished by that warmth of imagery characteristic of Oriental writings. But such a defense is not called for. The Christian believer may confidently challenge the scientist to show just cause for discrediting the statement. Indeed the late Professor O. M. Mitchell did not hesitate to assert that the rotary motion of the earth could be completely arrested in a few minutes without a single thing upon it being disturbed, and the arrest of this one motion was all that was required to effect the phenomenon. It would not have changed the earth's position in the heavens or its relations with its satellite, its sister planets, or its central sun. If the scientist can by his own will-power put out his hand and check the spinning of a top, what reason has he for thinking that God's will cannot check the whirl of a world? Has he any evidence that his will is more closely linked with matter than God's? The same eminent authority also pointed out that, if he had chosen, God could have lengthened the day by simply condensing the atmosphere and thus changing its power of refraction. Whether he actually adopted either of these methods or used a better we with our yet extremely meagre knowledge of nature have of course no means of determining, but we can see even now how such an end was within the ready reach of a will as masterful and as wise as we are warranted in believing God's to be.

The miracles of replenishing the widow's cruse of oil, turning the water to wine, feeding the five thousand with the five loaves and few fishes, though they involve something more than simply the overmastering of one force by another, as in the incident just cited, and are at first more difficult of apprehension and belief, and lie more exposed to the adverse criticism of scientists, yet, after a careful scrutiny, will be found, after all, remarkably analogous in many respects to achievements of the human will, and no more contranatural, or improbable, or wrapped in a profounder mystery. There is no necessity for thinking that in these or kindred acts any new matter or force was brought into existence. The oil and the wine, the miraculously provided cakes and fishes, differed in no respect in their elemental atoms, or in the combinations of these atoms, from products which nature, assisted and guided by man, had for centuries before been manufacturing. There was no call for any new matter, as it was already at hand in vast abundance. Christians need not claim this. Indeed, neither need they claim that, when, as it is recorded, in the beginning God created the heavens and the earth, he brought forth something out of nothing, as too many unthinkingly believe. Scientists may well pronounce such a notion absurb. An achievement like that would transcend even divine power, for it involves a contradiction, an impossibility. Something cannot come out of nothing. It is nowhere revealed that there ever was a time when matter did not exist. The beginning spoken of in Genesis need have reference only to the present order of things, the present processes of evolution through which the burning and non-burning balls of matter have been made to people space. Although this history may reach back over what to us are inconceivable periods, yet there unquestionably was a time when not a single sun or satellite anywhere existed, when matter must have been in some other radically different form. Further than this we need not go. If it was not originally a part of God, and is not now to be considered as an *emanation* from him, it must in our thought take rank as an equally self-existent and eternal entity. The fact is, the more prolonged and profound our study into its nature, the more impenetrable appears the mystery that shrouds it, for at first we can little realize that the substance we see and taste and handle is revealed to us simply by the effect produced upon our sense-nerves by forces that lie hidden behind it, so that we, when further advanced in our reflections, are led to query whether, after all, it is not the presence of *force* that is revealed to our consciousness rather than that of *matter* as the medium of force, and whether it is not of the existence simply of *force* that we have any certain knowledge.

As I have said, we need not infer that in these miracle-workings any new substance was brought into being, but only new methods adopted, or hitherto unused forces liberated, or greater direct power employed by a sovereign will in carrying out its decrees. The human will had before this accomplished the same ends in other ways, for how else can we explain the presence of the oil which the prophet found in the widow's cruse, or the wine already drunk at the wedding feast, or the bread and fish in the baskets of Christ's disciples before he miraculously multiplied them? But the human will had been compelled to resort to tedious and, for the most part, indirect methods to accomplish what the divine will wrought without delay, and apparently by direct impressment. I say "apparently," for it is quite possible that the methods employed were still indirect, though not accompanied with any noticeable delay. We ourselves are continually shortening the processes we employ in carrying out our purposes. By a more perfect knowledge of nature's laws we become more complete masters of her forces. What giant strides have we already made in this direction, especially during this nineteenth century! It is difficult for us to realize the nature and extent of our recent victories over matter. With what blank amazement would Washington and his companions be filled were they now, without knowing what had taken place, to return to the country they fought to save! For since Washington closed his eyes to earth, there have come the steamship, the locomotive, the telegraph, the telephone, the phonograph, and thousands of shortening processes. In his day, yes and forty years later, to cross the American continent was a task of many weary months. Now we make the trip in less than a week. The news of Waterloo was three days reaching England, but the tidings of the last bombardment of Alexandria, though halfway round the globe, took only as many minutes. The thunder of the first gun had hardly died away along the banks of the Nile before the air was throbbing with its echo on the banks of the Thames. We have also of late, through our telephones, succeeded in holding easy converse with each other, though separated by leagues of distance, even in actually distinguishing the peculiar intonations of each other's voices. At what time these discoveries of new forces and how to unfetter them shall reach their limit, who would be bold enough to predict? and yet not until science has won its final triumph over nature should devotees of science be unwilling to concede that it is clearly possible that Bible miracles were the work of Nature's forces simply guided by a will thoroughly conversant with Nature's laws, which were within the reach of the directive power of the will of a man if illumined by the insight of a God. But even if these miracles were performed by direct will-power, still we can point to constantly recurring instances in which precisely analogous effects are produced both in the vegetable and animal kingdoms, as well as in the higher realm of the human will. Scientific treatises call our attention not only to an inorganic, but also to an organic, chemistry, and assure us that the vital forces, working through complex animal and vegetable organisms, effect combinations of elements which outside of their laboratories or the laboratories of man are never produced, and are marked by extreme instability, readily decomposing under the influence of heat or fermentation, so soon as their influence is withdrawn. Those mysterious forces lodged inside the walls of seeds prove themselves the masters of other forces equally mysterious lodged inside the walls of atoms. Carbon, hydrogen, oxygen, and nitrogen never would have congregated into such chemical groups, or arranged themselves along such lines of symmetry, or climbed to such dizzy heights, directly against the steady pull of gravity, were they not working under compulsion; and so soon as they escape from the thrall of their taskmasters, their old individuality comes back to them, their old modes of combining, their old circles of association return, and the unstable organic compounds are torn down into the more stable, original, inorganic ones. Here we witness one great class of Nature's forces—the atomic—lorded over for a time by another and superior class. As we are daily witnesses of these facts, we never think of questioning them.

Further than that, we see the products of vegetive-vital forces taken possession of by animal-vital, and grouped into still more strange and higher compounds, and the chemic compelled to play a part still more foreign to their first estate. We know that this, too, is a case of compulsion, for the very moment vitality ceases, disintegration begins. These nitrogenous combinations are the very embodiment of instability.

We are daily witnesses of more startling wonders still.

They form part of our personal experiences. We find that we can by sheer will-power compel even these higher forces of animal vitality, and through them the lower, to do our bidding. The late Dr. Carpenter, the foremost physiologist of his day, called especial attention to this fact, asserting that thus we can greatly add to the acuteness of any of our bodily senses, can actually compel the nourishing blood to flow to any part of the system and infuse new vigor. The experiences of artisans and artists, astronomers and microscopists, experts and specialists in every class of work, deaf-mutes and the blind, abundantly confirm this. There are few of us who have not found by actual experience that by calling up certain thoughts we can turn the cheek pale or crimson it with blushes, flood the eyes with tears or make them merrily twinkle or flash with angry fire, cause the heart to violently throb or intermit its beats, throw the blood to the brain, make the knees quake, the skin perspire, the whole body tremble with intensity of emotion. The control which persons of cultivated histrionic powers have over the body to make it the vehicle of thought can be appreciated only by those who have witnessed the masters as they have entranced their audiences, and who have themselves been thrilled and spirit-bound under the spell of their enchantments.

If the vegetative forces can thus dominate over the atomic, the animal over the vegetative, and the will of man over all, what valid objection can science urge to the Christian's creed that God's will can by direct impressment effect combinations in the elements which Nature's forces indirectly and uncompelled bring about by slower processes according to the terms of their divine commission? Why may not God's will have as immediate and complete a sovereignty over the earth or the universe, as we over these complicate bodies of ours, which our spirits permeate through and through by their informing presence? And why may not his sovereignty be inconceivably more immediate and complete, and still retain in its relationships its marked analogy to the characteristics of force which science has herself recorded? Why may not the divine will not only make bread, wine, and fish directly out of the surrounding elements, but heal lepers, restore the blind, or even raise the dead, and still do no more violence to Nature's systems of law than the human will is doing every day? There are multitudes of well-authenticated instances in which persons have by simple determination checked for considerable periods the inroads of disease and even permanently broken its power. So startling have been the effects of the will and of the imagination over these susceptible bodies, there have arisen schools of theorists which advocate that what have hitherto been pronounced incurable diseases may be made to yield to the modern mind-cure treatment. They have doubtless overrated the will's curative energy, but they certainly have made no mistake, except in the extent to which such cure can be carried. Sudden fright, worriment in financial difficulties, brooding over loss of friends, remorse, chagrin, discouragement, loneliness, and longing,—all have their depressing effect on the body, and if not checked in time will lead to serious illness, if not to positive brain lesion. Glad surprise, large and unlooked-for success, the return of long absent loved ones, their rescue from danger or illness, appreciative sympathetic recognition of merit, fruition of long-deferred hopes, the stir of patriotic or religious fervor,—all have their medicinal influence, their exhilarating, uplifting power. Thoughts sudden and startling have often brought sickness or banished it, brought death even in the midst of healthful life, or lengthened life's lease for those apparently passing within the shadow. If impalpable thought is clothed with such recuperative and destructive power, and if between the Creator and his creatures there are open avenues of communication as there evidently must be,—avenues more open and numerous than between man and man,—what valid objection can be urged to the belief that God, with his infinitude of knowledge of the structure of the human frame and the laws regulating its processes, and with his intimate and accurate acquaintance with its ever-varying environment, can by turning the currents of thought by means of timely suggestions, by firing the fancy, rousing the conscience, raising the hope, occasioning and confirming the purpose, and, by the even more mighty magnetism of such positive and such sympathetic personality as his must be, summon health or sickness, life or death, when and where he chooses?

Thus the Christian's creed that God can answer prayer if he so desires, if in his wisdom it seems best, that there are multitudinous ways in which he may indirectly or directly carry out the mandates of his will without destroying any force or abrogating any law, finds in the discoveries of modern science most abundant confirmatory and illustrative facts. It is only in the ill-founded theories and misinterpretations of some of the devotees of science that its claims have been denied. Christianity will some day summon science to the bar of the world's judgment as her strongest witness and most helpful ally.

II.

But, query our doubting Thomases, suppose you can thus show that scientific discoveries warrant a belief in the possibility of God's effectively interfering in the course of nature and in the affairs of men, have they not also suggested and finally confirmed the opinion that, in point of fact, he never has; that, from the very first, matter contained the promise and the potency of all life; that the world is simply an immense organism which has reached its present complex perfectness through inherent forces working under fixed laws of evolution; that the stages of its growth have been as regular and predetermined as those of a tree; that its social amenities, its arts and literatures, its ripened civilizations, have finally evolved out of the original amorphic fire-mist through precisely the same regular gradations of growth as those out of which the rich grape-cluster or the golden-sphered russet has come to crown the long energizings of the germ-force that at the first lay hidden within the walls of the seed? We return to this query a most decided negative answer, and will endeavor to establish, as the second point in our present argument, that God has actually interfered again and again; that his interferences have not been confined to any one age, but have been present in all ages; that his will, by its creating and modifying power, has extended to all classes of phenomena; that his mandates are still being issued; and that their results, as asserted by recognized leaders in philosophy and in science, are present with us to-day.

At the first, matter was formless, motionless, forceless, structureless, rayless. On this there is now no controversy among the different schools of thought. Moses and Herbert Spencer, the creationist and the evolutionist, the dates of whose writing are separated by three thousand years, on this point clasp hands.

The belief is also as universal that this absolute simplicity of form and of nature has, after the lapse of ages, been converted into an almost infinite complexity, and that the cardinal changes have occurred in a certain order of sequence; but in answering the question as to how these changes have been effected, these schools of thought at once part company.

Those who affirm that in this unfolding there are no evidences of the active presence of an intelligent personal will-power are confronted by seemingly insuperable objections which science itself has furnished. Science discloses a law of inertia so far-reaching that not a single particle of matter in all the wide universe can set itself in motion. It also discloses that there is not a single particle that is now at rest. Whence that mighty initial impulse that thrilled through space and is still felt after the lapse of untold ages peopling the heavens with whirling worlds? Science also discloses that matter is made up of sixty-four or more different kinds of atoms, each inclosing within its walls, as we have already remarked, a force peculiar to itself, working under absolutely fixed conditions which no skilled chemist has ever succeeded in dislodging, or destroying, or changing in the minutest particular; each having all the characteristics of a manufactured article as affirmed by Herschel, Faraday, and Clerk Maxwell, and removed completely beyond the reach of nature's power or man's device to make or mar, alter or destroy. Out of these, through their mathematically exact chemical combinations, the whole inorganic world has been built. If there was once a time, as every evolutionist not only concedes, but stoutly contends, when every atom was precisely like every other, and not a single one had the faintest touch of attractive or repellent or affinitive force, through what instrumentality in some far past did these elemental forces, these individualized somethings, find birth and an abiding place within infinitesimal and indestructible walls of matter? We find on them no traces of development and no marks of decay. They are none other than God's immortals. Over the nature of their being, as well as over the cradle of their birth, there has been thrown a veil of mystery through whose closely woven meshes there comes no ray of revealing light to the anxiously peering eyes of science, and whose hiding folds no hand on earth has power to lift, except the reverent hand of faith.

Skilled specialists, after repeated trials to demonstrate that vitality may spring through spontaneous generation from dead matter, now candidly confess that all their efforts have thus far proved unavailing. Dr. Bastian with tireless zeal has worked to this end, and thought he reached it, but in every one of his experiments there has been detected some fatal flaw. The declaration that no life springs except from some living germ has stood the crucial test of the science of this nineteenth century. The lamented Agassiz affirmed this in his last lecture. Carpenter, Huxley, Tyndall, all the leading scientists, with refreshing candor, reaffirm it to-day.

With equal unanimity the world's savants point us to a fire period during which not only all the oceans and the soils, but the very beds of oceans, all the mines of metal and quarries of rock that form the earth, were once but drifting clouds of burning ether in whose fierce heat the hardiest germ would instantly shrivel and disintegrate. Whence, then, those first eggs out of which sprang the progenitors of those countless multitudes of living organisms that have from age to age so peopled our planet?

The secret of the egg, its nature and its origin, quite as seriously puzzles and confounds the evolutionist as does that of the elemental atom. Within its walls there hides a wonder-working fairy. Though not secure from intrusion, as is the oxygen or the carbon force, she as successfully eludes the prying eyes of mortals and is wrapped in as deep a mystery as to what she is or whence she came. With the lenses and mirrors of his microscope, the scientist tries to look through the curtained windows of her palace. Baffled in that he presumes with subtile chemistry to bolt unbidden into her very presence, but the sprite, warned by the first footfall of the intruder, passes with viewless feet through some secret postern gate out into the unknown beyond, and never comes back again. After this he compounds in his laboratory the like chemical ingredients of which he has found the egg composed, and in precisely the same proportions, and then exposes this, his skillfully built protoplasm, to a carefully adjusted heat. Weeks pass, but no life. For a third time he finds himself facing failure. At last, with humbled pride, he accepts the truth that germinal force is not some property inherent in matter, but rather an organizing impulse introduced from without, separable at any time from the mass over which for a season it is made dominant, the product of a personal creative will whose impalpable thought it is commissioned to incarnate into living form.

Again, between not only the four primordial divisions of the animal kingdom and also the classes, orders, and genera, but even the one hundred and thirty thousand different species, it has been demonstrated, after a century of most painstaking exploration and experiment, there have been great gulfs fixed which no natural, delegated force has power to pass. Within certain lines it has been discovered that species can be modified into varieties through climatic or dietetic influences or cross-breeding, but changes thus effected are found quite unstable, the parental types reappearing through the law of atavism when in new surroundings or removed from the culturing care of man. But, however, when an attempt is made to develop absolutely new, distinct species out of old ones, naturalists encounter in the law of the sterility of hybrids an uplifted iron hand, and hear a stern voice, saying, "Thus far, but no farther." That voice they are rapidly learning to recognize as the commanding voice of God.

The origin of bodily organs is another of nature's many secrets to which evolution theories furnish no key. These organs are found on examination to be contrivances of the most complicated character, combining often into a single group hundreds of closely correlated parts so nicely adjusted, so absolutely interdependent in many instances, that the absence of any one would not only seriously cripple the others, but render them totally inoperative, hopelessly defeating the purpose of the mechanism. These parts being thus unquestionably complemental one to the other and incapable of performing any useful office unless combined, their origin and present combination can be accounted for only as a projection into physical fact of an ideal previously conceived and matured by some organizing mind. It seems absurd to suppose that each part could have been originated independently, without any reference to the others, and slowly developed, in its own time and way, out of some minute, indefinite, fortuitous variations, either through the influence of its environment or through some internal blind force, into its present perfected and permanent form, and then that they all, through some chance circumstance, should have fallen into each other's company, and have proved so exactly suited and so absolutely essential each to each as to become at last thus inseparably associated in close corporate work.

Exploring parties of geologists, naturalists, and anatomists, after having with inexhaustible patience, with trained powers of observation, with most ingenious instruments of research, ransacked the rock record of earth's crust down through even the Silurian strata to the very dawn of being, and having examined the present occupants of every continent and sea, now testify in the name of science that nowhere among extinct species or living ones have there come to light any facts proving that there ever were any such processes as evolutionists so boldly announce to have taken place in introducing the different gradations of sentient life on this planet.

The same is true of the many curious instances of mimicries in nature, and indeed of all phenomena of correlated growth.

Materialistic expounders of the universe also find themselves confronted on every side by the ever recurring phenomena of instinct and are at their wits' end to account for that perfect poise and mastery of body exhibited by some animals directly after birth, for that accurate intuitive knowledge of perspective, that minute familiarity with first witnessed scenes, that unrivaled ingenuity of design and flawless finish in mechanical execution of works performed without experience or a guiding model or the aid of instruction, that instantaneous grasp of the most occult principles in natural philosophy and profound acquaintance with the laws of chemical and vital action, and especially that far glance of prophecy on the accuracy of which depend the lives not only of individuals, but even of entire species. Theorists who cling to a naturalistic explanation denominate instinct a lapsed intelligence, affirming that it is the accumulated wisdom of past generations acquired through painful and protracted experience and handed down under the laws of heredity in the form of fixed habits and of constitutional mental bent. But scientific investigations in natural history have brought to light thousands of facts to which such an explanation is wholly inapplicable, which fairly laugh these theorists down.

The spider that builds its tiny diving-bell, anchors it with strong cable to the river bottom, and distends its walls with air pressed from entangling meshes of web on its abdomen, and then, with this, its royal pavilion, that shines through the water like a globe of woven silver, rears with watchful wisdom, amid seemingly most hostile surroundings, its brood of hungry children, is one out of a vast multitude of living witnesses that testify to a direct divine informing of the mental life below the human, the impulsive promptings of instinct being followed blindly by those creatures which stand thus in imperative need of its guiding wisdom. As well accredit an intelligent self-conscious purpose to those particles of matter which, when the time is ripe, arrange themselves with such promptness and precision along the lines of symmetry which form the faces of crystals or the exquisite patterns of flowers, as to ascribe to these lower orders of sentient being the knowledge, the invention, and the prescience which their works display.

But over the question of the advent and distinctive attributes of man the battle of the schools has been most hotly contested, calling into action on both sides every reserved force of scholarship and mental acumen, as the issues at stake transcend every other, involving not only the foundations of theistic faith, but even the very evidences of an endless life.

The extensive scientific investigations which have grown out of this heated controversy have brought to light a vast array of most interesting and significant facts to which the extreme evolutionist and the equally extreme creationist have both gone for corroborative proofs of their theories, and neither of them gone in vain.

Man in his body, in his instincts, and in his mental traits, bears many very striking resemblances to the brute tribes, suggesting some closer tie than the strict creationist is yet ready to admit; although out of the lines of affinity with the numerous ape and lemuroid species that are by scientists classed with man in the sub-orders of primates, there could be constructed, as a distinguished writer has remarked, "only a net-work and not a ladder." There have also been found in man equally marked differences, suggesting, on the other hand, that in effecting the changes there were actively present higher forces than mechanical or chemical or even vital, and that there was introduced, as in the case of the atom and the egg, an absolutely new ingredient, of which there was no *germ* even, anywhere existing.

In man we miss the brute's great teeth and claws, we note fewer instincts, a lessened speed, a weakened muscle, a blunted sense, a back laid bare, a skin left tender; divergencies which would denote marked degeneracy were they not most strangely accompanied by a vastly increased mass and multiplied convolution of brain. Here appears that same deep correlation on which the parts of a bodily organ are built, bearing the same emphatic testimony to the prior existence, the personal presence, and the plastic power of some intelligent, organizing will. To be sure, there is here no change in the material ingredients. Neither is there any, when out of the soil a flower unfolds its tinted petals and fills the air with its fragrance; but as the soil, the moisture, and the sunlight have no power to thus combine into this marvel of grace and color and sweetness until the directive force of some buried germ thrills them with its talismanic touch, so neither in the body of the brute nor in the nature of its environment dwells there any power known to science capable of producing such a circle of complemental changes, physical and vital, as mark the advent of man.

Furthermore, science in its explorations in the higher realm of thought has brought to light a class of phenomena so entirely novel as to indicate that there has taken place something more than a mere modification of the four forces, mechanic, atomic, vital, and instinctive, which have been successively set at work in the world, that an absolutely new force has been ushered in, a force possessing characteristics so fundamentally different from all others that they can in no sense be regarded as its progenitors, and a force not only of a uniqueness so complete as to thus preclude any suggestion of kinship, but of a uniqueness so peculiar that it becomes a travesty on scientific interpretation to explain it simply as an unfolding under the universal law of evolution of another one of the hidden, inherent properties of matter. And this new force, known as a self-conscious and a responsibly sovereign *ego*, is apparently the exclusive inheritance of man, is his distinctive feature, lifts him completely up out of the low plane of brute being.

In the mental life below the human there are semblances of self-conscious, deliberative thought, of moral discernment and of responsible free-will; and instances of this nature are so many and so striking, the belief is prevalent, not only in scientific but even in religious circles, that we differ from the brutes only in having a clearer thought, a deeper discernment, a wider freedom; but there are now advanced investigators of the highest attainments and of international celebrity who believe that those semblances are wholly delusive, and that in this mysterious pantomimic life below us there are no really reliable evidences of the presence of a distinctive, self-conscious, spiritual force constituting true personality. Animals unquestionably possess in common with us blind instinct, sensation, direct perception, association of objects and ideas, automatic attention, involuntary memory, indeliberate volition, reproductive imagination, sympathetic emotion and emotional expression. Nearly, if not quite all of the phenomena of their thought-life can come through the exercise of just these low forms of mentality and do not necessarily imply that they ever get beyond the domain of the senses, that they have any abstract, deliberative, introspective thought, that their consciousness ever reaches up into consciousness of self. Their mental states may be, and probably are, simply passive; their memories and imaginations but prolonging and multiplying their sense-perception through laws of association and suggestion.

It is true there are some few phenomena that do not seem susceptible of this explanation, but as we find clearly within the charmed circle of instinct, where there is uniformly nothing but blind obedience to a God-given impulse, acts which to ordinary observers show deliberation, design, profound reasoning, even moral purpose on the part of the animal, we naturally feel warranted in assuming that these occasional instances met with apparently outside of this circle, and indicating that animals at times really enter within the vestibule, at least, of self-conscious life, are delusive, that the real mental background to these unvoiced acts may after all be God's, and not theirs.

The belief that thus with the advent of man there was introduced an entirely new force, a spiritual, self-conscious, personal entity, seems to find further warrant in the fact that he alone has ever manifested a desire or shown a capacity for progress, intentionally improving on the past. Did animals really have souls in them, did they possess truly reflective faculties like our own, the developing influences of the tens of thousands of years, that have one by one rolled round since their life began, would have wrought in them an advancement so marked that their mental status would long since have been placed beyond all controversy.

That this non-progressiveness is not rightly chargeable to bodily imperfections is clearly evinced in the wonder-workings of the ant, the spider, and the bee. Apes have hands but they have never yet built a fire or replenished one, or shaped a tool.

This belief finds still further warrant in the fact that with brutes instinct reigns; with man, reason; that they have their thinking done for them, he is forced to do his himself; that they reach perfection, without effort, at a single bound; he, if at all, only after repeated and disheartening failure; that with them the final purpose seems to be simply to conserve the body, with him, to improve the mind; that with them the supplying of physical wants brings unbroken peace, a deep content, the horizon of their thought shutting closely down about the now and the near; with him there is ever a vague unrest, an unsatisfied longing, an indefinable dread, angel-winged expectancies.

How can we account for God's pouring out such wealth of inventive thought in care for brutes' bodies and showing not the least concern, so far as we can see, for preserving and developing anything nobler, except on the ground that he has planted in them no germs of anything nobler to be developed, that he has never given them any real, personal self to be conscious of, that with them body is absolutely the very top of being?

While then there are strong suggestions, if not positive evidences in nature of some mysterious relationship between men and brutes, that relationship is certainly, as I have already suggested, as remote as that existing between the flower and the soil out of which it springs. The dull clod has no magic gift of self-transfiguration but displays merely a capacity for a passive yielding to the plastic touch of some newly arrived vital force, when out of its well-nigh shapeless, scentless, colorless dust are wrought the queenly robes and peerless perfume and richly crimson blush of roses.

The investigations of science bring the certain knowledge of the direct action of the divine will still closer to us, even within the circle of our own individual experiences. Sir George Mivart, Fellow of the Royal Society, who stands in the forefront of science, and Professor Rudolf Schmid, President of the Theological Seminary at Schonthal, Würtemburg, who stands in the forefront of philosophy, claim that self-conscious and responsibly free spirits must be new and independent existences transcending nature, they going so far as to state outright that each human soul is the result of a separate creative fiat of the Almighty.

We might enforce this their position by remarking that out of the old nothing new can come except new combinations, and the soul is believed to be an absolutely new element and not simply a new form of an old one. This our self-consciousness positively affirms, and we must implicitly rely on its testimony or our whole foundation for any belief is hopelessly swept away. It also says that each soul is an indivisible unit, that there cannot be transmitted from parent to child any portion of the *ego*. Resemblances may be, but nothing of the child's spiritual entity has been or can be derived from his progenitors. Human souls are God's direct gift. To the fashioning of each one he has given his personal attention. It is only its fleshly covering and its other material environment he has entrusted to the care of secondary causes.

Facts brought to light by modern scientific investigations and closely analyzed by modern scientific methods, are thus daily diffusing and deepening the belief among the candid and thoughtful that the progress through the ages from the simple to the complex, from amorphic matter to a peopled world, has been something more than a methodic, self-originated, and self-sustained evolution of elements held hidden in matter from all eternity, that absolutely new forces have from time to time been introduced from without through direct creative fiats of a personal will, the old forces, inside their limitations being, as the work progressed, utilized, when found available, simply as avenues for ushering in the new.

III.

WE now come to the third general division of our theme, that God not only can effectively interfere, either by direct or indirect methods, without working any disorder, abrogating any law, or destroying any force; and that he not only has, in fact, thus interfered again and again in all ages and in countless matters of moment, but, further, that it is not only not presumptuous, but most natural and reasonable, for us to expect that he will interfere for us individually, however insignificant we may at present seem to be.

It is claimed by those who controvert this position, that God has, as we have already remarked, adopted broad, comprehensive plans, in which he has regard to general interests, and not to exceptional cases; that in these plans he is as unyielding as granite; that his interferences have been in the nature of creative fiats, simply for completing these wide-reaching original designs; that he has no time or thought for individual cases; and that, if any one of us would secure any of the benefits of the present order, we must make these plans a careful study, and adjust ourselves to them as best we can, and not expect their author to break in upon them and give his personal attention to our private, insignificant interests. In other words, we must rely on our own exertions for any modifications of our environments, must master the secrets of nature, comply with her laws, if we would make her forces our servitors and become masters of our circumstances.

There is apparent warrant for such a view. It would seem as if the individual were indeed lost sight of,— everything is on so vast a scale, every part of this wonderful mechanism of a world is so far-reaching in its results. The earth's whirl on its axis brings day and night for all; the inclination of its axis to the plane of its orbit and its circuit round the sun determine the change of seasons, the rise and fall of tides, the width of zones, the force and direction of the great trade-winds, the character and limitations of vegetable growths, the nature and habitat of the fishes, the birds, and the beasts. The sun ceaselessly pours out in every direction that mysterious influence which we call light. It indifferently enters hovels and marble halls. It comes through every open doorway, every uncurtained window, every crack and crevice. It purples the velvet petal of the violet and fills it with fragrance, and afterward, with seemingly heartless haste, rots that same petal to shapeless, colorless, odorless dust again. It kisses the sheltered valley into waving harvests, and at the same time, with other of its rays, scorches the sand wastes with death's desolation and silence. At one time it darts in through the pupil of the eye, and with exquisite art transfers to the retina the outer glory and thrills the soul with strange rapture; at another, when the delicate nerves are aflame with fever, it tortures with its touch, and blisters and blackens that very same canvas it had with its swift pencil painted with splendor. An atmosphere miles in thickness completely envelops the earth. It forces itself in everywhere. All gills and spiracles and lungs must breathe it, though sometimes it comes loaded with poison, instead of balm. Now with gentlest zephyr-touch it gratefully fans the cheek of an invalid, anon with the swift sweep of a cyclone it levels a forest or unroofs a city. Water is as omnipresent as air. The air is indeed permeated with it, as all substances, fluids and solids, have their every particle encased in air. What interminable leagues of tossing billows, with their glistening foam-caps breaking over the white-winged sea-gulls of commerce as they hasten on venturesome errands over the treacherous depths, some to reach safe shelter, it may be, in distant ports, some to fly wildly before an angry storm and sink into the opening jaws of a hungry sea! Fire, though not actually, yet potentially, is also omnipresent. Even the ingredients of water itself will burn, and in the fierce flame which their chemical union kindles, the metals and the earths, even fire-clay itself, will be consumed to ashes. Forests, grasses, and peat bogs, underlying beds of coal, countless reservoirs of oil, are ready for the torch. Angels and demons of combustion are all about us. They stand in waiting on every hand, ready with their ruddy faces to beam kindliest cheer from our furnaces and chimney corners and swinging chandeliers or to blaze in mad fury amid the crumbling walls and rafters of our homes. They will cook for our tables, smelt our ores, draw our trains of trade, turn the wheels in our workshops, multiply our comforts a thousandfold, or, if we are not aware, will, as very fiends in their wild work of a night, turn our proud Chicagos into smoldering ruins. In some far past the whole earth was but a burning ball, and lava streams and earthquakes and smoking craters tell us that the primal fires still rage within. This elemental force has been provided on a grand scale. The economic scheme of which it forms a part embraces the farthest fixed star in its infinitude of thought.

Electricity, the latest utilized force of nature, has been found to bear the same stamp of universality and to stand toward us in this same twofold relationship. It falls from the clouds in death-dealing thunderbolts; it also with deft fingers renders invaluable service in the civilizing arts of life. It becomes the winged Mercury of the mind, carrying thought-messages across continents and under seas with well-nigh the swiftness of light.

As we thus study nature force by force, attribute by attribute, and note this feature of universality pervading all, and this dual relationship which each sustains of blessing or cursing, as angel or devil, how powerful and painful the questioning, whether, after all, it is not too true that exceptional cases, or individuals during exceptional crises, have failed to enter as factors into the thought of God in the dispensations of his providence; whether individuals have not been placed in the midst of the same possibilities; and whether it does not rest with each to bravely make the best of his environment, and trust to his own right arm and stout heart to carry him through! And, besides, is not God's universe so wide, are not his cares so multitudinous and complex, that he has time to make only general classifications, establish wide-reaching laws, delegate great secondary causes, arrange his forces on a scale graduated with mathematical precision, and set them at work in grooves unalterably fixed? Is he not necessitated to take simply a sweeping glance, to contemplate in the mass the swarming myriads of beings evolved from the dust as the grand processes of life go on? Has he not thought it sufficient to establish the great dynasties of organized living creatures that through the ages have seemed to rise and sink with the regularity of the tides of the sea? We cannot even number the massive worlds which he has set whirling through illimitable space, and which must demand at least his general supervision and require his constantly sustaining power.

At first glance we are apt to conclude, viewing the subject from this standpoint, that there is indeed no individualizing in God's providences, no attention paid to detail, no more note taken of the units that make up the mass than the farmer takes of the separate kernels of wheat which he harvests from his fields. Here moves by a cloud of locusts dense enough to darken the sun; an east wind rises and greedy ocean-billows swallow them up. A volcano bursts, and a Herculaneum with its thronging human life is swiftly buried in a grave of ashes. There comes an earthquake shock, and a Sodom sinks into the sea; a steamboat disaster, a railroad accident, a visitation of cholera, a breaking out of fire, a caving in of a colliery, a whirl of a cyclone, and scores and hundreds of human lives perish in an hour. Is it probable that the individual arrests the attention of the Almighty in the great ongoings of his providence? Have you and I, in our little corner, ever attracted his attention, much more excited his interest? Has his great heart ever beat in love for each one of us? Has he ever called us by some dear name and watched with tender solicitude the unfolding of our powers, entered into sympathy when our hearts have bled with bereavement, or been crushed with failure, or made desolate by estrangement or unfeeling neglect? How many hours in the life history of every one of us are darkened by a sense of utter loneliness! How many times our hearts cry out for the appreciative sympathy of a divine companionship! Oh for that comforting assurance which blessed Christ's sorrow-wrung heart when he said, "And yet I am not alone, for the Father is with me"! Is it presumptuous for us to think that that assurance may also be ours? That it is not, I believe to be the unmistakable teachings, not only of the Sacred Scriptures, but of all animate and inanimate nature and of all sound philosophy.

The Scriptures are full of this consoling revelation. There is rarely a page not illumined by it. To teach it was one of the distinctive features of Christ's ministry. How he delighted to dwell on the brooding watchfulness of the Father! In reassuring his disciples he told them that God, who gave his personal attention to the clothing of the grass and the lilies, and was not so great or so busy as to overlook the fall of even a little sparrow, surely would keep loving and sleepless watch over them. Even the hairs of their heads, he confidently assured them, were all numbered.

Such like disclosures, so many and so explicit, throughout the books of the Bible, find most abundant confirmation in the facts of science. The geologist and the chemist, the botanist and the naturalist, have in their separate departments found phenomena which the Christian philosopher may boldly claim as incontestable evidences of God's sympathetic presence with his children. The more deeply nature is searched, the more convincing the proofs of God's infinite painstaking for his creatures. His plans to these ends have evidently been thought out in their minutest details. We are overwhelmed with astonishment as we see into what small concerns he has suffered his thoughts to enter, and out of them by an ingenuity of contriving possible only to a creator of limitless resources has wrought results of far-reaching import. No candid student of nature can fail of becoming profoundly convinced that there is absolutely nothing, however inconspicuous, that does not only embody a divine thought, but in some way plays a part in carrying out the promptings of a divine love.

If any one in his hours of depression is haunted with the feeling that he is too insignificant to attract God's personal attention, much more be the object of his constant loving care, he will find himself wonderfully reassured if he will lay down the telescope and take up the microscope, for he will soon see that the fault is all in himself, in that he has had a far too meagre conception of God's thought-range and breadth of sympathy. Such an examination will disclose to him that, as a positive fact, God has somehow found abundant time, notwithstanding the multiplicity and the magnitude of the interests of his vast universe, to give his personal attention to the equipping and provisioning of beings of infinitesimal minuteness. That mighty hand in whose hollow the heavens are held, has also sufficient delicacy and precision of touch to fashion the finely reticulated wing of the ephemeron. The same art-conception and marvelous skill that paint the sunset and bend the rainbow have touched with most brilliant pigment each feather in the plumage of the fly. The same musician who has also conceived the grand organ harmonies of ocean billow and thunderburst, has also adjusted, part to part, with loving care, that sweetest of musical instruments, the throat of the skylark, whose wild rapture of song so thrilled the ethereally gifted Shelley that he immortalized it in verse as the blithe spirit-voice of the air.

God apparently shows not only the same infinitude of care, but the same keen personal delight, in his works in the domain of the minute as in that of the vast and the mighty. Look deeply as we may into nature with our most powerful artificial lenses, even to the very microscope-limit, we can detect no hasty oversight, no cold indifference, but exhaustlessness of patience and lavishment of thought, and in every detail of each work an absolute faultlessness of finish. Illustrations of these comforting truths abound all about us. The world is full of them, but I have time to cite only two or three.

There is a class of microscopic animals, the Diatomaceæ, which have existed in such vast numbers that entire mountains have been found composed of their remains. The forms of their infinitesimal shells when magnified are discovered to be of most exquisite beauty and of every conceivable pattern. "In the same drop of moisture there may be some dozen or twenty forms, each with its own distinctive pattern, all as constant as they are distinctive, yet all having apparently the same habits and without any perceptible difference of function." Neither sexual nor natural selection has, so far as we can discover, any governing influence here. In these varied beauties are there not evidences, which scientific theorists have so far failed successfully to controvert, of God's giving his personal attention to the adornment of the minutest of his creatures, to his conceiving and embodying in innumerable faultless forms and pleasing combinations of tints his conceptions of beauty? How this infinite painstaking has benefited these mysterious specks of life, we have no means of determining. Perhaps they come and go without having the faintest intimation of the symmetries and colorings which the Divine Architect and Artist has, by the interposition of direct will power, introduced into their calcareous palace homes. We cannot prove that it was for their especial benefit these patterns and paintings were designed. Perhaps the ultimate purpose was the æsthetic culture of inquiring human souls, or it may be that other and even higher ends will come to light in some after age. Certain it is such painstaking implies a purpose, and whether we can discover it or not, the fact brings with it, to every thoughtful mind, with overwhelmingly convincing force, that God is personally conversant with, and has taken an active personal interest in, the life-furnishings of creatures so minute that their individual forms are to us absolutely invisible without the aid of the microscope, and so low in the scale of being that naturalists are still divided in opinion as to whether they are animals or plants.

The inorganic world equally abounds in illustrative proofs of this same comforting truth. I will select a single one. The luminous flame that has brightened human homes through all civilized centuries is an aëriform chemical combination of hydrogen with oxygen and carbon. The difference in the degree of inflammability of the first two gases is the cause of all the illuminating properties of the flame, and yet that difference is so slight that the times of their ignition are separated by a period absolutely imperceptible to our unaided senses. The hydrogen takes fire a very small fraction of a second before the carbon, and as it unites with the oxygen of the air it lets go its chemical hold on the carbon, which the instant it is thus released changes from a gas to a solid, so that into the colorless flame of hydrogen is constantly being showered the finest carbonic dust. These minute particles become little glowing coals emitting a brilliant light just for an instant, and then, like the hydrogen, spring into the chemical embrace of the all-devouring oxygen. The infinite painstaking here displayed, the delicate nicety of adjustment, the critical attention to the minutest details, are no less astounding than the world-embracing beneficence of the results.

The case of the little brown water-spider, to which brief allusion has already been made, is the only other illustration I shall have space to give of God's personal, painstaking care over the minutest matters in his kingdom. In common with the numerous species of this order of articulates which abound in all parts of the world, this diminutive creature has had given to it four pairs of seven-jointed legs, the last joint armed with two hooks toothed like a comb, frontal poison-fed claws, eight eyes and a multitude of spinnerets from whose infinitesimal openings issues a glutinous liquid which the instant the air strikes it hardens into threads invisible from their fineness until they are massed together into a single, strong, elastic cable. But it has furnishings and instinctive impulses peculiarly its own. Its body has a thick covering of hair which it has been taught to most curiously utilize. Strange to say, this air-breathing animal is prompted to build its home and rear its little ones on the beds of streams, and the devices by which it has been enabled to surmount what to us would seem insuperable obstacles may well fill us with admiring wonder. It weaves a diving-bell, air-tight, mouth downward, and ties it tightly to the bottom. Then coming to the surface it covers its hairy abdomen with fine web, lies on its back until all the interstices between the hairs and the meshes of web are filled with air, swims under the bell, presses out into it the entangled air, comes again to the surface, and repeats the process, until all the water at first in the bell has been displaced, and the bell made habitable.

In all this procedure the spider has unquestionably been guided by him who equipped it. No candid and appreciative observer can fail to note this, for what, can it be imagined, first determined it, supposing it to be following out its own thinking, thus to locate its nest under water, for it has no gills fitting it for such a habitat, or how did it study out so ingenious a method for making such an undertaking possible? The inventor of this bell must have known that air is lighter than water, that it can be mechanically retained in fine fabrics, and that when introduced into an inverted receiver it will crowd out the water, instead of being absorbed by it. Has this spider been so close a student of nature as to have discovered these laws of physics, and is it so gifted an inventor as thus ingeniously to have applied its knowledge, without either instruction or experience? This daintiest of palaces must have been thought out in all its details before the spider began spinning its first thread, for the weaver shows no hesitancy and makes no mistake. It must also have been the work of a single mind, for its parts are so intimately correlated that the absence of a single one would not simply obscure the conception, it would totally destroy it. There must be either perfection or flat failure. This alternative was presented to the first spider of the species. I would like to show, had I time, how this little creature is also equally blessed with divine guidance as to how and where it shall deposit its eggs, how enwrap them in clusters with silken cocoons for protection and warmth, when and how to release the tiny babies from their coverings and transport and feed them when first they come, as they are sure to do, in swarming and hungry companies.

The equally marvelous prescience and skill displayed by all instinct guided creatures and their equally marvelous equipment for their work, afford us illustrative proofs without number of God's most intimate acquaintance with, and loving care for, the momentary interests of earth's speechless, soulless, perishing myriads. Neither their implements nor their skill can be accounted for as the slow outcome of stern experience, for their instinctive promptings are followed blindly, and their wisdom and skill antedate experience, and are independent of the aids of instruction or of any working model. To the progenitors at least of every animal species, there has come a direct divine impressment and informing. New wants with correspondingly new implements and new instinctive impulses issued from the creative will of the Almighty. Provision was doubtless made at the incoming of each species for the transmission, through laws of heredity, of such traits as should constitute its distinctive endowment, and thus a general supervision over each species instituted.

But still more specific provision seems to have been made to cover exceptional necessities, to answer the demands of exceptional crises in the individual lives of the seemingly most insignificant. There appears to have been left a certain latitude of modification and amendment of instinctive promptings. As I have already remarked, animals unquestionably possess, in common with us, not only blindly followed instincts, but sense-perception, association of objects and ideas, automatic attention, involuntary memory, indeliberate volition, reproductive imagination, sympathetic emotion, and emotional expression. Though the phenomena of their thought-life may be classed under these lower forms of mentality, though they may never rise to deliberative, abstract, introvertive thinking, may never attain to self-consciousness, having no self to be conscious of, may never have the clear light of reason or ever exercise a responsibly free choice, yet they do seem to have had some means provided for supplementing instinct in those peculiar emergencies for which no general provision through instinct could be secured. This clearly evidences to us that God's providential care, even over the lowliest, extends beyond the segregated mass that constitutes the species to each separate individual in it, and even to that individual's exceptional needs. The thinking here displayed, though outside the circle of instinct proper, will still be found, on final analysis, to be God's, and not theirs.

To receive the full force of this comforting truth, we must keep in mind that all this loving care is taken for creatures of a day, who are here hemmed in by simple sense, and who have promise of no tomorrow; and we must also keep in mind, what science has not only conclusively demonstrated, but illumined and glorified by its extensive researches, that man is a microcosm, the crown of creation, the consummate flower of all the ages, that it was for him this world was provided with its mineral deposits, rock-quarries, and coal beds, with its vast reservoirs of oil, its dense forests and waving grains and grasses, with its flocks and herds, with its mighty elemental forces, with its flower-petals, its arching rainbows, and its painted skies.

It was to secure for him, nature's sceptered king, a fitting environment, that all the mighty processes of evolution had been carried on through all the untold geologic eons of forgotten time, and it was for him earth was fitted up, not as a permanent home, as the all-in-all of his existence, but simply as a first year's training school for powers which, though barely budding now, have in them the promise and the potency of an endless life and of a divine likeness. A single deathless human soul outweighs in worth ten thousand worlds of lower sentient life.

Having described at some length, in a paper entitled "Science and Christ," the discoveries and conclusions of science as to man's place in nature, and having no space here for its general discussion, I will content myself with the simple statement that the more profoundly phenomena have been studied by scientists and scientific philosophers, the more clearly and gloriously have shone out the truths to which I have just alluded; that God has been busied through untold ages in preparing for man's advent, that man has been the grand goal of his endeavor, the *ultima Thule* of his creative thought on this planet; that all this prolonged preparation could not have been merely to render comfortable a short-lived and low-planed animal existence, that this patient approach could not have been to a consummation so inconsequential and unworthy, but that he for whom the centuries have been so long waiting and to whose coming they have been pointing with prophetic finger, who fulfills the types, completes the prophecies, wears the crown, surely was not born to die; and that he who has proved himself capable of unraveling the intricacies and following the vast sweep of the divine thought as is evidenced by his discoveries in science, his classifications of knowledge, his advancement in the arts, his rapidly approaching universal mastery and ingenious utilization of nature's forces, his unconscious duplicating of God's thought-processes as incorporated in the lives of the world's silent, instinct-guided workers and in the mechanism of their bodies; he who has proved himself capable of so apprehending the spirit of God's vast creative plans as to be able to become his sub-creator, noticeably multiplying and improving the products of vegetable and animal life, making the waters swarm, turning deserts into gardens, developing the crude possibilities of untamed nature; he whose whole being can thrill with harmonies of sound, of form, and of color, and who has not only reproduced them but carried them to grand exaltations in oratorio and sculptured marble, speaking canvas, cathedral pile, and landscape gardening, and has laid all matter and even all force under tribute to his æsthetic tastes; he who can thus enter with keen appreciative zest and assimilative capacity into the thought-life of God; and, finally, he who has had entrusted to him, what far transcend everything beside, the responsible gifts of moral discernment and liberty of choice, out of which alone character can come, surely must have reached, in point of privilege, the very top of being, and must possess in living germ the very attributes of God himself, with all the golden possibilities of growth in God's eternal years.

When we thus attempt to measure the worth and dignity of man, we must also keep in mind that each individual soul comes fresh from the Creator, and is not simply the product of processes of evolution begun in some far age and perpetuated by secondary causes which God has long since ceased to superintend and to whose general outcome alone he has ever directed attention. The soul's environment, its body and its wider surroundings are, indeed, the result of such processes, but each soul is in itself a unique spiritual entity, bearing the imprint of a distinct personal purpose, and constituting the embodiment of some cherished ideal, some fond anticipation, some sacred love, right out of the very throbbing heart of God.

The drift of the centuries has been to an ever more complete development of individuality; it has been a progress from homogeneity to heterogeneity; such has been the history of evolution from the dawn of time, as Spencer, Huxley, and thinkers of that school have, through learned and brilliant treatises, informed the world.

It is not the great mass as such that excites God's loving interest, but the individualized units in it. It was not the creating and provisioning of a mighty human race simply as such that was the *ultima Thule* of his thought, but the developing of the distinctive personal traits of individual souls, and the establishing with them at the last, after discipline has done its work, intimate and eternal companionship. To think that God ever purposed to stop short of this would be to belittle his plan, belie the teachings of all sound science and philosophy, leave the grand scheme of evolution incomplete, and judge of God as being coldly self-contained, craving no sympathy, contentedly sitting apart in eternal isolation, wholly unresponsive to the tender pleadings of his children.

When we discover that God has given his personal attention and poured out a wealth of inventive thought on every particle of dust, on every minutest fiber of every leaflet, on every organ of every infinitesimal creature, we can no longer reasonably withhold our faith in his sympathetic presence with the humblest of his human children. And so science will eventually forever silence the fear of the self-depreciating, who, in their discouragement, are tempted to doubt whether the great God of the universe has ever in the vast multiplicity of his affairs particularly noticed them, much more kept loving and tireless watch over their personal destiny, or ever sought for their confidence and the outpouring of their longing and their love.

But science has not only convinced us that we have no valid reason for questioning God's sympathetic presence, but furnished the strongest possible grounds for resting our full faith upon it, and making it the delight and inspiration of our burdened souls. Those grounds it furnished the moment it published its discovery that every form of vegetative and animal life demanded an environment, that it has no resources in itself for self-maintenance, and that also within its reach it invariably found that on which it was fitted to feed. Plants have required soils and sunlight and distilling dews, and they have found them. Though almost countless the peculiarities of need, no species has appeared for which provision has not been made awaiting its advent. The seaweed found its ocean bed and salted surf; the cactus, its parched sand plain; the lichen, its rock; the edelweiss, its Alpine height; the gills and fins of fish, oceans of water; the wings and lungs of birds, oceans of air. Our eyes have found objects without to be painted on their retinæ within and artist-sunbeams to paint them; our olfactories, the air loaded with odorous exhalations; our nerves of taste, a wide variety of flavors to select and enjoy; our ears, all nature vocal with a grand concert of song. Not only are our bodies constituted to touch and take in an environment and find one wondrously suited to every need, but the same is true of both our intellectual and emotional capacities. All nature abounds with suggestive thought. It is full of mental stimulant. It is a book in which every grade of intellect finds passages of absorbing interest and deepest import. Its leaves are turned eagerly by prattling children, gray-haired *savants*, matter-of-fact men of affairs, dream-enamored poets, and system-building philosophers. Its lore is still unexhausted, though the human race for scores of centuries has sought to master it. It has depths of meaning which human insight has not yet fathomed: heights of sublime exaltation to which not even the most spiritually gifted have yet attained. It is full of open letters to every son and daughter of earth with every sentence penned by a divine hand. Our longings for intellectual and sympathetic interchange with our fellows have been met through literature and arts and architecture, through family ties and ever widening social circles. But with this almost infinite painstaking to provide a fitting environment for man, there is a want which in all the fulness of God's works there is absolutely nothing suited to satisfy. Man in his higher nature craves a sympathy which no creature can give. Unless these spiritual aspirations and deep longings, the sure tokens not only of his divine sonship but of his divine likeness, can find a divine environment of companionship, of interchange of thought and affection, all that is God-like within him will languish and die and he sink to brute life or below it. National and individual history, wherever people have self-exiled themselves from the Father, has furnished sad cumulative proofs of this. Is it reasonable to suppose that a plan so wonderful in its elaborate painstaking and masterful achievements, exhibiting such seeming exhaustlessness of inventive resource, would fail just where a failure must prove so disastrous. Is it reasonable to suppose that God would create man with a capacity and a longing for his own sympathetic presence, indeed make that presence necessary to his well-being, and then withhold it; that he would give him spiritual lungs on whose respiration of an atmosphere of divine loving recognition his spiritual life depended, and then leave him to pant and die in a vacuum? These questions carry with them their own emphatic denial. To proclaim this grand fact of God's sympathetic presence and to embody it in a life was the glory of Christ's mission to this sin-cursed and sorrow-burdened world. He even sealed it with his blood.

Thus from nature, philosophy, and the revealed word there comes to this life-giving fact a threefold confirmation.

In our lonely hours, in hours of desperate battling with temptation, of bitter bereavement, of perplexed and care-cumbered thought, at times when our hearts bleed with poignant regret or through unjust accusation, when friends on whom we have leaned or in whom we have confided the sacred secrets of our inner selves have become estranged, through the long days of languishment on sick beds, in moments when with streaming eyes and trembling lips we bid good-by to loved ones, in every hour of need, we are privileged to say, as did the Saviour when the dark clouds gathered about him; "And yet I am not alone, for the Father is with me."

Out from God's sympathetic presence into the chill night of an endless death the incorrigibly wicked finally go away. Into it the lovingly obedient come, into its welcoming smile, its golden sunlight, its eternal day.

IV.

I HAVE thus far attempted to show—

1. How God can interfere in nature whenever he chooses without working any confusion, abrogating any law, or destroying any force;

2. That he has thus actually interfered, and that repeatedly;

3. That we are, each one of us, of sufficient importance to warrant his interfering for us.

I now desire to consider whether we can reasonably believe that he will interfere because we ask him, doing for us what otherwise he would not have done.

In following out the different lines of inquiry suggested by this theme, we have found the whole earth instinct with the Divine Presence, that whichever way we turn we stand face to face with nature's God, witnessing not only finished works replete with his thought, but works still being carried on by organized and tireless living forces. These forces we have found wrapped in such unfathomable mystery, working right before our very eyes with such unabated vigor, such wondrous precision, such wisdom, such irresistibleness of movement, that we have recognized divine thought and divine power in every bit of rock crystal, every pendent leaf, every tint of sky or painted petal, every liquid note of bird, or restless tongue of flame. And it has greatly enhanced our pleasure to find that our own minds are so akin to the divine that we can trace with clear, interpretive insight the great trend of God's thoughts through the ages as they have become incarnated one by one; for when, from off that illumined face confronting us everywhere, there thus fades that strange far-away look and in its stead comes an answering glance of recognition and kindly greeting, that face apparently draws so near we can all but feel its warm touch upon our cheek, look down into the infinite depths of its love-lit eyes, and see the parting of its lips as they break the long-kept silence with words of benediction.

But it appearing that these forces are derivative and delegated, rather than direct acts of divine will, we have found that we must take other steps in our thinking before we can reach that assurance for which every human heart hungers, of God's still being present on this earth and still actively interested in it; for otherwise, what grounds have we for believing that these forces were not fully commissioned ages ago, and that since then God has gone far into the stellar depths to people other planets and never once come back again or even given this little globe a passing thought? for otherwise, how do we know but that the earth is nothing more than a finished piece of mechanism, like the watches we carry, and, like them, wound up and kept running by the coiled energy of some hidden spiral spring? Happily we have discovered that matter and force are of such a nature, and so related, that abundant opportunity has been afforded, and with apparent design, for the effective intervention at any time of direct will-power. A study of our own experiences has suggested this; for, if we by the might of our own wills have wrought such multitudinous changes on the earth, we can readily conceive that the divine will can work by analogous methods, and be as much more effective as the divine knowledge transcends the human. It cannot, as we have found, be reasonably urged that this, God's, direct personal interference would be a confession of flaw in his scheme of evolution, as provision for this may have been, and doubtless was, a part of that very scheme. He, as we have seen, left many of his works incomplete with the evident design that man's will should complete them; and if provision was thus made for the after use of the guiding force of the human will, why not for that of the divine? And we are confirmed in this faith when we reflect that, otherwise, God, instead of being an exhaustless fountain of outflowing, energizing thought, instead of being to us the very personification of living force, of tireless mental buoyancy and zest, becomes a picture of changeless, thoughtless, emotionless calm, of absolute mental stagnation; all the vast plans of his whole universe of worlds, having been inconceivable ages ago, not only determined upon to their minutest details, but intrusted for their unfolding to agencies fully commissioned and empowered to carry out those details to the very letter. Since that time, which lies in a past so remote that no finite imagination can conceive it, he must have been lying with folded hands and folded thought and folded feeling, virtually dead in the midst of the abounding life which he himself created. This conception of the divine existence is repellent to every earnest active soul, and there is nothing in the discoveries of science to compel such a belief. The perfecting of the intellectual and spiritual in man must, of course, be God's highest work here, and command his chief attention. But he has linked the soul indissolubly with matter and cosmic force in this world certainly, and also in the next, if the Bible disclosures be true; for after death our souls, so says the record, will still be clothed upon, though the garments be of an imperishable and glorified texture. So we have no warrant in affirming that God has withdrawn his personal oversight and interference from any, even the lowest of his kingdoms, so long as they are so inseparably intertwined, and exercise over each other an influence so vital and lasting.

The facts of the past as disclosed by science, we have found to confirm us in this faith; the progressive changes from a first formless chaos of dead atoms to whirling sun clusters and solar systems of organized peopled worlds being but the stately steppings of a creating God, and testifying to a sleepless watch and tireless activity as the ages have one by one rolled by. On this revelation of God's mode of existence in the past we may safely predicate that of to-day and of all coming time. We can feel assured that his hands will never fold in weariness in caring for his own, that his eyes will never close in listless inattention to their fate, that he will never surrender to delegated forces the full conduct of the complex affairs of his universe; but will ever be a commanding and directing power everywhere present to the uttermost bounds of space,—just as the vital forces within the boundaries of these bodies of ours sway the cosmic, only more perfectly; and as our spirits, so mysteriously housed within, order the organs to answer the behests of their all-governing wills.

But having progressed thus far in our attempted solution of this most perplexing problem, we find ourselves confronted by questions far more formidable than any we have yet met. They are questions which are sure to intrude whenever there is any thorough thinking on this theme. They have proved such fruitful sources of doubt in earnestly inquiring minds, that, instead of being, as they too often are, ignored or evaded by the leaders of Christian thought, they should be squarely met and fully answered. I remember stating them once at a prayer-meeting presided over by my pastor, who was also a college professor; and, although they were perfectly germane to the subject of the evening, and I asked for light and needed it, he simply remarked, "There is some intellectual difficulty in that," and immediately passed to other things, and neither in public nor private discourse did he in the slightest manner ever again allude to them. This reverend teacher in his evasive indifference is, I fear, far from being an exceptional case, for it has never been my fortune to have either heard from the pulpit or seen in print any attempted reply.

Grant, says the doubting Thomas, that it is true and demonstrable, as claimed, that God can interfere, that he has interfered and is still interfering, and interfering every day and hour, in every individual life, watching that life with loving interest and with unremitting care, still what proof is there, in all this, that prayer has in a single instance effected any change in the plans which God had formed before the prayer was uttered? Has any prayer given God any new information as to the needs of any petitioner; or rather, has not God had from the first an infinitely fuller and more accurate knowledge of the entire life-necessities of every soul than the soul itself can ever possibly have, with its imperfect finite faculties and meagre experience? Is it not absurd to imagine that we can in any way instruct Jehovah? Do not our prayers appear to him who knows our real needs but utterances of wildest absurdities? But suppose they do sometimes actually voice our real wants, have not those wants already been known to God and definitely provided for by him? Has he not been busy for ages fitting up this world for us? Are not those instances of his direct interference which are insisted on as having actually occurred and as still occurring, as much parts of this original plan as the formation of a crystal or the growth of a tree? Has he not thought out to the minutest detail just what to do and how to do it? Are the forces at work in the world, and their combinations, so complex that exigencies are constantly arising which escaped God's foreknowledge or for which he failed to provide? Does science or revelation afford us any warrant in thus limiting God's wisdom or questioning the perfection of his works? If God thus thought out deliberately and fully his vast plans before he uttered his first creative fiat, and had as his guide a perfect and all-comprehending foreknowledge, think you his will has since become so vacillating that he can be cajoled against his best judgment, or that more kindly feelings can be enkindled within him, by the blind, passionate pleadings of creatures of his own make, and whose lives are yet but in the bud?

The only reply I have ever heard given leaves the difficulties just where it found them. It is this, that the prayers of God's people have been all foreknown to him, and their answers provided for, uncomputed ages before they were uttered; that they entered into God's thought when he formed his original plan, and were made to constitute an integral part of it. This reply is so plausible and has given such general satisfaction, that it may be regarded as the accepted creed of Christendom.

Suppose this were true, that God has both foreknown all prayers and made ample provision for each as each deserves, would not the difficulties just urged still remain? For if the prayer of a righteous man availeth much, as the Scriptures teach, and if it had influence with God, as Christians believe, what matters it, so far as these objections lie, whether that influence is exerted now or was exerted ages ago? For, according to the supposition, prayer has actually wrought a change in the divine purpose just the same, only at an earlier date; and it is just as truly an embodiment of the blind longings of a finite being addressed to an infinite God; and the fact of the prayer's availing—which must mean, if it means anything, that it actually effects a change in God's plan at the time its influence is felt—witnesses just as pointedly against the perfection of God's plan, since it existed before the change was wrought, and against the stability of his purpose, whether that change occurs now or took place before the chaotic fire-mist was rolled into suns. But, say you, how, then, can the objection be answered? Only in this one way,—by denying the doubter's major premise, that God's foreknowledge is all-comprehending. The denial of this, I believe, can be shown to be in perfect consonance both with sound philosophy and the revealed word when once that word is rightly understood. Let us then examine this denial, first, from a philosophical standpoint, from the standpoint of the science of metaphysics.

If God foreknows everything that will ever come to pass, all his own mental states must necessarily be included in that foreknowledge. His eternal past and eternal future must be to him an eternal now. This is axiomatic. A moment's reflection will convince us that otherwise there is not a single present intention or plan but what is exposed to the possibility of modification. If a single thought or emotion is ever going to spring up in God's mind unanticipated, coming in as a complete surprise, God himself must be as ignorant as we as to what part of his vast plans it will pertain, or what will be its relative importance, or what the radius or duration of its influence. Indeed, both radius and duration must be absolutely infinite; for, however minute the influence or modification, it must result in others, and those in others still—the circle widening thus without end; for the parts of God's plan are supposed to be intimately interlinked, complemental, so precisely fitted part to part that the effect of each is felt throughout the whole, like the intricate complications of a piece of mechanism. And if one thought or emotion may thus spring into being unanticipated, be absolutely original, why not ten or ten thousand? Indeed, what limit can be placed on their number or on their modifying power? And so, if we would logically defend a belief in the all-comprehensiveness of God's foreknowledge, we must affirm that not a single new idea can arise in his mind—not a single new emotion be felt, and that if he is thus limited now he must have been equally so at every moment in all the eternal past, and must be through all the years to come; for if there ever has been, or ever will be, a moment when a new thought can thus come, then during all the time preceding that moment the foreknowledge was incomplete. Where does this lead us? In what sort of an intellectual or emotional condition does this irrefragable logic compel us to assert God to be continually? Unquestionably that of perfect stagnation. No thought processes can be carried on under such conditions—no succession of ideas, no change of mental state; but God must have been, and must still be, imprisoned in a hopelessly dead calm.

When then did he form his plans for creation? Under this supposition, there never could have been a time when he began to think about them, nor a period during which he adjusted their different parts, each to each, in that perfection of harmony which so astounds us; for that would involve thought-succession. We are not at liberty under this supposition to affirm even that the entire plan in all its details flashed instantly upon him,—for this would impeach the perfection of his foreknowledge up to the instant of such inflooding of thought, but must content ourselves with asserting that it has existed in his mind from all eternity as one of its constituent elements. If God has had no thought-succession, he can have had no feeling; his emotional state having ever necessarily been that of unbroken placidity—of absolute apathy, his heart throbless as stone. He could experience no change of feeling; for that would involve thought-succession. From all the sources of joy or sorrow of which we can conceive, he would be utterly debarred—from pleasurable or painful memories, from hopes and forebodings, from social sympathies, from emotions that accompany changes, contrasts, surprises, from the glow of activity, even from the delights and griefs of contemplation; for they all involve thought-movement. Therefore under this supposition God can have no emotional activity, for he would have no thought-activity for its background. Thoughts must course, must come and go, or the heart lies dead.

Such are the absurdities in which we become hopelessly entangled the moment we attempt to defend the doctrine of God's perfect foreknowledge. And besides, on further reflection, we will discover that it is, after all, utterly impossible, from the very nature of the case, for God to foreknow all his own future. The very fact that he is a sovereign spirit precludes this. It is equally impossible, and for the same reason, for him to know what our future will be. He has made us equally with himself of sovereign will, and placed upon us all the responsibilities of that sovereignty. When he thus created us in his own image, he, by that very act, surrendered a part both of his power and of his foreknowledge. He has left it possible for us, despite all the influences he can bring to bear, to rebel against his throne and persist in that rebellion. He in thus constituting us the arbiters of our destinies, necessarily circumscribed his own power. There was no other course open to him. We must be free, must be sovereign, if we become morally accountable, and ever reach up out of a state of simple innocency to that of divine virtue. And God when he thus surrendered absolute control, also of necessity limited his foreknowledge, for our own self-study reveals that our perfect freedom of choice is inseparably linked with uncertainty as to what that choice will be. Character can be evolved only out of struggle. Virtues are the names of victories won over temptations; and where temptations environ a sovereign will, there must be risks, a certain degree of uncertainty. It cannot be otherwise. We cannot exercise this sovereignty or know that we have it, unless there are open to us two or more courses from which to choose, and our fidelity to principle or the depth of our self-sacrificing affection cannot be developed or brought to test except by genuine wage of battle. And how can it be certainly known whether this shall issue in defeat or be made glorious by decisive victory? From the very nature of things, complete foreknowledge is precluded, for we can go in the direction of either the weaker or the stronger motive. But, say you, perhaps we have the power thus to go, but in point of fact we never do, for the motive that controls us proves itself the stronger in that we invariably yield to it. This is too wide a conclusion for the premises. Our yielding does not prove it the stronger intrinsically, but simply relatively, and then only because we make it so through our choosing to direct and hold the current of our thoughts in that direction until the chosen object of contemplation acquires prominence and power. We cannot stop the flow of thought, but can change its direction. And even God himself cannot with unerring certainty predict what that change will be, for it is purely an act of sovereignty. If, in fact, we never go in the direction of the weaker motive, how do we know we can? Would not this unbroken regularity prove the presence of inexorable law? The testimony of our inner consciousness that we could do differently, would under such circumstances never come to proof. And yet only where strict regularity prevails, can the necessary data be obtained for perfect foreknowledge. Outside this circle of responsible sovereignty, under the reign of absolutism, of immutable order, within which the physical and vital forces and the pure animal instincts work their wonders, God can of course predict with unerring certainty, and to the minutest detail; for the plan is all his own, and from it there is not the slightest deviation, nor can there be. Courses here are predetermined and as exact as mathematical formulas. God, who fixed the conditions, who founded the laws, must know the issue. But in the region of delegated sovereignty, of absolute freedom of choice, of moral accountability, uncertainty just as necessarily enters in and renders prediction impossible.

If what I have argued be true, we need no longer struggle with those hopeless tasks of harmonizing foreordination with free will, and of explaining how a beneficent God could bring into being souls which he at that very time positively knew would be eternally lost.

The doctrine of God's perfect foreknowledge is not only unphilosophical, but also unscriptural. The Bible exhorts us to the deepest earnestness in prayer,—to downright importunity,—and encourages us to believe that the fervent prayer of the righteous man availeth much. No petitioner can plead with any genuine unction unless he believes that he can actually effect some change in the purposes existing in the divine mind at the time his prayer is offered. If he were convinced that everything had been prearranged from all eternity; that his tears, and sighs, and passionate words of longing had been present in God's mind always; that they never had exerted, and never could exert, any influence, effect any change, as there could never be a time when they would first arrest God's attention,—how could he wrestle, agonize, in prayer? It would seem but empty show to him, that he was merely playing a part. Every word he uttered would fall back dead. If he believes in God's foreknowledge, he must, while he prays, if he prays as the Bible commands, utterly forget his belief and fall into the temporary delusion that the matter is yet undetermined, that God's heart is tender, can be moved, that his purposes can be changed. He must forget his belief, must go ahead just as if foreknowledge were not true. Think you God would force his children to such straits, to such mental stultification? The thought is repellent. Read if you will the ninth chapter of Deuteronomy. Moses here rehearses the several rebellions of Israel, and his three separate pleadings before the Lord, of forty days and forty nights each, without either eating bread or drinking water. Each time he fell down before a very angry God who had fully purposed, and had definitely announced his purpose to destroy the rebels, and each time, if Moses can be credited, he actually changed that purpose right then and there and rescued his people. The God here depicted had none of that foreknowledge which theologians with such strange unanimity ascribe to him. But, say you, that and similar accounts scattered throughout the Bible are simply instances of anthropomorphism, of rhetorical accommodation, of describing in the language of human experiences and human limitations what really transcends the human; that it was not the intent to have these narrations interpreted as literal history, but as poetic approximations or dim shadowings of really ineffable truths. It seems to me that it would be a strange way to bring the truth within our comprehension, to state what is directly opposed to the truth, and to reiterate the downright falsehood, again and again, in a most misleading way, and in a matter of such vital moment that all possibility of religious life depends on it, and through which alone any lasting comfort comes to the hungry human soul. Could Moses have thought that what he was so importunately pleading for had actually been determined upon millions of ages before, and that the picture of his prostrate form, his streaming eyes, his starving body, his passion-swayed soul, had been lying in the divine mind from all eternity? He unquestionably believed directly the opposite, and the narration was designed to teach us that directly the opposite was true.

Think you that Christ during that long night of agony in Gethsemane, when he cried out over and over again, while great drops of blood stood on his brow, "If it be possible, let this cup pass from me," knew all the time that there was but one way in which the race could be rescued, that precisely this one had been predetermined to its minutest detail, and that all that was left for him was to carry it out to the bitter end? Were not those the agonized utterings of a faithful yet shrinking human soul,—for Christ was human as well as divine,— poured out before a supposed loving and sympathetic Father? And have we not a right to believe that they not only deepened God's sympathy, but actually influenced him to again reconsider the whole subject, that happily he might discover some escape for his Son from the impending doom? When Christ prayed, he unquestionably meant the same as if he had directly said, "Father, do think it over again, and see if it be possible, and if it is, let the cup pass," for the petition is pointless unless this thought is embodied in it. Christ had not yet for an instant harbored the thought of relinquishing the enterprise or even imperiling it by any attempt at self-rescue. He did not even ask for sustaining grace. All he plead for was another more searching inquiry to see if some different means of rescue could not be devised. He simply desired to avoid needless humiliation and pain. In what a pitiable farce he must have consented to become an actor during the watches of that memorable night, if he positively knew all the time that there was no other way possible! And if he did not thus know, but God did,—and that too from all eternity, even to the precise mode and to its every detail,—and had unalterably determined upon its being carried out to the very letter, with what cold, relentless cruelty this Father must have listened, hour after hour, to that sorrow-stricken Son as he plead in heart-rending agony for him to see if there were not some other equally effective way to save the lost! How could he listen to that pleading, wailed out on the night air, for something he had not the faintest idea of granting? Why did he not encircle him in the arms of his everlasting love and at once explain the impossibility of change, if he certainly knew that no change was possible? What importunate pleading! No parallel can be found in all human history. Was it for naught? Was it a stupendous blunder born of ignorance? We cannot mistake it for some blind outcry of a sinking soul. Should we not seek for some sane, sensible purpose in the plea? We have here revealed not simply one of the disciplinary seasons in Christ's career, his desperate battling with the tempter, for he had betrayed no weakness, no unwillingness to face, if need be, any fate however terrible. He showed from first to last a spirit of perfect submission, for note how carefully he coupled with his passionate prayer, "Not as I will, but as thou wilt." Nothing could be added to his consecration. His self-surrender stood complete. His soul was white as the light that beats on God's throne. But how natural, and necessary, and full of deepest significance, appears this whole scene in this, earth's darkest tragedy, the moment that we conceive that Christ, instead of being crazed by his grief, was quickened by it to clearer spiritual insight; that in his cry, "O my Father, if it be possible, let this cup pass from me," the real plea was that the whole subject-matter of modes of rescue should be reopened and again most searchingly reviewed; that God fully answered that prayer by a long, deep study; and that, when the last faint ray of hope went out in night, he in accents tender as an infinite pity could make them, told Christ all; and then the Saviour, satisfied, rose from his knees, wiped away the blood-stains of his agony, and with a calm, majestic bearing—that never again left him, save in the last throes of dissolution—said to his disciples, "Rise up, let us go; lo! he that betrayeth me is at hand."

Had I time, and were it necessary, I might multiply indefinitely citations from Scripture of cases in which it is clearly taught that even to God's eye the future is not wholly uncurtained,—that he carries on processes of thought as we do, elaborates plans, modifies them and sometimes even abandons them altogether to meet the demands of unforeseen exigencies as they arise, that he interferes in behalf of his children and because they ask him, actually forming and executing entirely new, unpremeditated purposes in response to their asking.

Against this view, that we actually exert an influence over the divine mind, it has been urged, as I have already remarked, that it implies imperfection in the divine adjustments, and vacillation in the divine will, that it is the very height of presumption in us to suppose that we can influence the great God of the universe to do differently from what he had in his wisdom deliberately planned. The usual reply, that God has from the first foreknown all prayers and carefully incorporated his answers into his original designs, is, as I have endeavored to point out, fatally lacking both in sound philosophy and in Scripture support. How, then, can the objection be met? In the first place, God has, as I have explained, left his works in such plastic state that he can whenever he chooses interfere by direct will-power without occasioning any disorder. If so, what can be urged against the belief that he left them thus with the express design of introducing from time to time such modifications as circumstances should require? Indeed, what other explanation can be given than this for the presence of this universal characteristic? This, instead of betraying a weakness, a flaw, in God's plans, reveals its strength and finish. So far as it was possible for him to perfectly foreknow, so far the conditions of change and activity have been unalterably fixed, as in the operation of chemic, vital, and instinct forces. But realizing that in delegating to his human offspring the responsible power of free choice he would necessarily let in the element of uncertainty, thus obscuring his prophetic vision, he with most profound wisdom contrived through this very plasticity in nature to be able to meet any emergency that might arise, to leave every avenue free, every particle of matter and every form of force promptly responsive to his call. His plans in such a case, instead of being ill advised and marred with faults, are simply unperfected and in constant process of completion. He is thus afforded ample opportunity to enjoy unceasing mental activity, and with sleepless eye and tireless hand to be ever caring for his own. To me this conception of God is by far the most exalted and stimulating. Instead of an idle spectator walled out of his own universe, he becomes an intense participant of effective personal presence, a living, loving spirit, free and masterful, the embodiment of all the active virtues and throbbing sympathies that are the necessary heroic belongings of him who would win the affectionate reverence of human hearts.

God being able to forecast the general trend, the ordinary tendencies, of the lives of his children, has unquestionably prearranged his providences to meet their probable wants, has provided for them a bountiful environment full of illimitable possibilities of joy and growth. For the extraordinary and unforeseen he has made provision by leaving himself ample facilities for immediate interference. And then, too, by timely suggestions he may, and often does, make us willing and intelligent servitors of his will, inaugurating by a single whispered thought, in moments of crisis, movements of deep and lasting import in our own or others' destiny.

Thoroughly conversant, as he must be, with all the peculiar mental states of every individual as fast as they arise, his seed-thoughts fall opportunely into responsive soils and soon quicken into harvests. A word dropped into the mind of a young Luther starts a reformation that shakes to its very center the papal throne of the world. As Carlyle says, "The clock strikes when there is a change from hour to hour, but there is no hammer in the horologue of time to peal through the universe when there is a change from era to era." God notes those pivotal periods and uses them.

Any human will obstinately standing in the way of the great ongoings of his providence, as it certainly can as long as it is free, he reserves the power of either temporarily or permanently placing under duress. Of course, while thus borne down by a superior personality, while deprived of its freedom of choice, it is relieved of responsibility, its acts lose their moral quality, and it becomes like any other force in nature. It is, however, responsible for necessitating such summary procedure. This divine impressment, this infringement upon our freedom, may, for aught we know, be frequently resorted to in the course of individual or national history. We certainly are the arbiters of our destinies. But woe betide him who recklessly dashes against the thick bosses of Jehovah's buckler. We are closely hedged in by carefully constructed systems of inexorable law. We can break those laws if we choose, but we do it at our peril. We can stand out persistently against all God's good influences; we may render futile his utmost efforts to rescue us from the thraldom of sin. The whole race may combine successfully to thwart his purposes of love. From the very nature of the case he was forced to incur that risk, for virtue can live only in an atmosphere of liberty. But we must remember God's unalterable determination from the beginning has been not to make everybody loyal and loving, but simply to furnish the possibilities for loyalty and love, and then do all in his power consistent with the conditions precedent to character-forming to develop within each soul the germs of divinity of his own hand's planting. He may be forced to summon a deluge, or an earthquake, or some wasting pestilence to do his terrible bidding; he may be forced to abandon what after trial prove ineffectual methods, and adopt new ones; he may be forced to recall the gift of liberty, or the very gift of existence here and hereafter from those who persistently repel all proffers and become hopelessly hardened; but his loving purpose still holds out, his laws still stand, the golden opportunities are still presented, each century witnesses some new conquests of love, some souls added to heaven's company, the great scheme is steadily going forward to its final glorious consummation.

Such a view of God—of his maturing and executing plans, of his intellectual and emotional life—as I have endeavored to present, is the only one, after all, actually conceivable by finite minds. To pronounce him unconditioned, unchangeable, omniscient, ommipotent, omnipresent, using these words in their ordinary and fullest acceptation, placing no restriction upon their meaning, is simply falling, unintentionally no doubt, into nothing less than word jugglery, affirming what to human minds must of necessity be absolutely unthinkable. The only rational course is to take for our basic thought that we have been created in God's image, and then to picture God as a spirit possessing in perfection attributes analogous to our own, although these are yet germinal and sin-distorted.

I am now ready to answer the question, How can we reasonably hope by our petitions to effect a change in the divine purposes, and why should we plead importunately, why kindle our souls into such intensity of fervor? The Scriptures in enjoining earnestness need not be understood as favoring attempts to coax and tease God, as we too frequently do our earthly parents, to act against his better judgment out of some weak, shortsighted sympathy. If that be our purpose, we may be certain of flat failure. Our prayers will never induce him to deal any more generously with us. He has always stood with outstretched arms, with overflowing sympathy, waiting impatiently to bless us. What untold wealth of deep inventive thought, what untold eons of slowly passing years he has already lavished in his preparations for our coming, for our maintenance, for our unfolding, for our permanent weal! While our prayers will not make him any more kindly disposed, will not noticeably increase his sympathy *for* us, they will in most marked measure increase his sympathy *with* us, will profoundly change our attitude toward him and multiply our capacity for blessing ten thousandfold. Indeed, so radical is the change wrought, that what would have been poison before, becomes medicine now. We thus furnish God new facts upon which to act, facts of mental attitude, the unforeseen outputs of our sovereignty. That attitude is one of Christ-like love, manifesting itself in five forms,—that of willing obedience, of self-sacrificing service, of sense of divine dependence, of restful confidence, and of intensest longing. Until that attitude is attained in all these its prime essentials, God, if he should interfere by stepping outside his general providence, in which the evil and the good are served alike, to confer especial favors, would be doing violence to his conceptions of fitness and of true beneficence, would work his children a most positive injury, placing a premium on qualities that stand over against these forms of love, thereby countenancing a spirit of rebellion, selfishness, self-sufficiency, distrust, and ignoble apathy. It is the fervent prayer of the righteous man that availeth much. He must be righteous and his righteousness must be on fire to fulfill the Scripture conditions. That availing power is something more than retroactive; it moves the arm that moves the world. As this is a moral state of the soul within the circle of its sovereignty, the product of its absolutely free choice, there cannot be, as I have shown, any sure prophecy of its coming. But when it comes, all barriers are burnt away. Reserve gives place to closest sympathetic intimacy. What more natural when the spirits of father and son thus meet and mingle, than that the son, care-cumbered it may be, or broken with grief, or baffled in purpose, though battling still, should pour out in most impassioned utterance his deep and noble longings? Love itself would so prompt; for love casteth out fear, is the very essence of liberty. Cautious reserve cannot live in its atmosphere of holy confidence. All curtains of concealment fall instantly at the magic touch of sympathy. He could not keep his longings back. His father's tender look and tone would break the seals of science, would touch his lips with coals of fire. The thought of trying by coaxing to melt down his stern reluctance is utterly foreign to such a scene, repugnant to such a state, and was never contemplated in the gospel. What more natural than that God's heart should be deeply stirred by the fervid outflow of such a passion of love and longing, and that he should by direct will-power supply the deficiencies of his general providence, or by timely suggestions reveal its resources, and place them in reach to meet the needs of such a soul in such an hour?

These views are not only thus in deep accord with the principles of sound philosophy and the revelations of modern science, but also with the profoundest intuitions of human hearts; for when once our sense of world-dependence and of self-sufficiency is rudely swept away by some disaster, and we come intently to long for what we find we cannot reach without God's help, how soon we brush aside all hindering creeds, and in dead earnest plead our case, and plead believing that the heart and arm of God will answer to our plea! But in this intensely materialistic and scientific age there have so insidiously settled about our thought the bewildering fogs of learned and subtile sophistries breathed out by those who would either relegate God altogether from his universe or make his relations quite inconsequential and remote, that only in the distressing stress of crises in our history do our long-neglected religious intuitions assume their rightful sovereignty, and restore us to our true relations with him who in his great love never wearies in caring for his own. But may we not hope that the night is well-nigh spent, that the fogs are lifting, that a new day dawns—a day of deeper, clearer, truer thought, of more perfect knowledge, of more enlightened faith, and a faith whose kindly light will prove the sure harbinger of God's perfect day?

V.

I HAVE thus far endeavored to show—

1. How God may interfere whenever he chooses;
2. That there are incontestable evidences, and multitudes of them along down the centuries, that he has thus actually interfered;
3. That we are warranted in believing that we, each one of us, the humblest and most obscure, are of sufficient consequence to attract his attention and secure this his direct interference; and
4. That he will interfere because we ask him, doing for us what otherwise he would not have done.

There is left for me now but one other general affirmation to make. With its explanation and proof I believe I shall have presented the subject in all its essential phases. It is this: Every reasonable prayer offered in a right spirit is certain of favorable answer. This is the clear import of Christ's comprehensive promise to his disciples, as recorded in Matthew, "All things whatsoever ye shall ask in prayer, believing, ye shall receive," or as Mark states it, "Whatsoever things ye desire, when ye pray, believe that ye receive them, and ye shall have them." If we interpret these passages in the light of the context and of the general trend of Christ's teachings, we cannot but conclude that Christ premised in his promise that the prayers should be reasonable and that they should be offered in the right spirit. No petitioner who complies with these two conditions need ever fear failure.

To have our prayers reasonable, we should, in the first place, guard against asking for anything which we can procure by our own exertions, making use of the resources of physical and mental strength, of social ties and general surroundings already in reach. God is a strict economist. If he has already made ample provisions in his general providence, and if we ourselves can by proper industry discover and utilize this provision, we ought not to expect from him any further help by special act. We must exhaust our own means first, and ask him simply to supplement our weakness and insufficiency. Otherwise we would be asking not only for what God has really already bestowed—and bestowed in a way which he thought would do us the greatest and most lasting good—but for what, if granted again in this more direct manner, would prove to us a positive bane, and not a blessing; and if such a course were continued, all incentive to industry and enterprise would thus be taken away, physical and mental sloth would succeed to healthful, growth-promoting activity, abject timidity and feeling of dependence would take the place of a manly spirit of self-reliance. No wise parent among us, however keen and quick his sympathies, would ever consent thus to shield his child from toil and care and battle test, for he knows he would by dandling him thus in the lap of ease and luxury be sure to unman him, weaken his body and invite disease, dull the edge of his faculties and rob him of every prospect of progress, of every trace of nobility, of everything that gives zest and incentive and joy to life and gilds the future with its pencilings of glory. Wise teachers refrain from helping their pupils so long as they can help themselves. Their office is not to relieve but to incite, not dwarf but draw out, not convert those under their charge into cowering weaklings but into athletes and conquerors. Even the eagle, prompted by a divine wisdom, will push her timid fledglings out from their lofty eyrie-home, and watch them flutter and hear their cry of distress as they disappear down the sides of the gorges; keeping herself, however, meantime, in ready reach, and now and then darting under to save them from fatal fall, for God has taught this mother thus to throw her children on their own resources, that they may feel their wings and learn to use them. This is a rude awakening. It seems a cruel banishment. But otherwise they would never learn to poise and wheel in air, to dart like thunderbolts, to breast the hurricane, or to climb the steep stairways of the sky.

God loves us too wisely and too well to heed any

of our cries except in times of positive and pressing need. He will let us struggle alone until our strength and judgment fail. He will, however, always keep in call, and will in deepest sympathy watch the contest point by point, and we can rest assured that in the hour of our extremity, should such hour come, we shall be made gladly conscious of some answering heartbeat, shall hear some whispered word, shall feel the uplifting power of some helping hand of love. A prayer for God to convert our impenitent friends would be unreasonable if without conditions or provisos, as it might be utterly impossible for him to secure such a result. All we can sensibly ask for is that he will make use of all the instrumentalities at his command, arrest the attention, rouse the conscience, reveal the danger of delay, the consequences of continued rebellion as well as of loving obedience,—in a word, bring to bear all the persuasive influences possible and still leave their wills untrammeled, for without absolute freedom of choice being constantly maintained, no moral change can possibly be wrought.

Again, our prayers to be reasonable must be consistent in all their parts, must be free from contradictory requests. To answer such prayers in their entirety would be impossible even to God. To illustrate: It would be inconsistent for us to ask only for the agreeable things of this life,—for freedom from care, sorrow, and pain—from disappointment, privation, calumny—

from all the vexations, perplexities, and disasters of life,—and at the same time that he would develop in us that glorious Christ-likeness for which in our nobler inspired moments we so intently long; as well ask for the knit sinews of an athlete, while nestling in undisturbed repose in the padded sleepy hollows of a rocking-chair. The ignoble fate of a soul set free from life's carking care and environed with all that the most cultured civilization could suggest, Tennyson in his "Palace of Art" has pictured with a master hand. If we would be like Christ, we must pass through Christ's school of experience. He needed the discipline of suffering and struggle, as well as we. He began where we begin—in perfect innocency yet characterless, possessing simply the possibilities of virtue totally undeveloped. It is because he afterward became a hero, battle-taught, battle-tested, battle-scarred, and yet never knew defeat; it is because he through faith wrought righteousness, out of weakness was made strong, endured the cross, despising the shame, suffered long and was kind, sought not his own, was not easily provoked, thought no evil, rejoiced not in iniquity but rejoiced in the truth, bore all things, believed all things, endured all things, loved us with a love that never failed and loved us to the end,—it is because of this, Christ has stood before the ages, and will stand, as the Peerless One, the Revelator of the Divine Heart, the Liberator and Saviour of mankind, the Prince of Peace. We must bear Christ's cross, would we wear his crown.

We fall into these contradictions in our prayers, through a total misconception of the design of this life. Evolution, not unalloyed present pleasure, is the purpose now. We have been housed in perishable bodies full of quivering nerves; have been environed with antagonistic forces that threaten and thwart us at every turn; our paths have been left rough and full of dangerous pitfalls; poisons pervade much of the air we breathe, the water we drink, the food we take to repair these weak clay tenements. To millions, life is a heavy care-burden, a fierce contest, and how frequently is it one long catastrophe made up of broken hopes and baffled purposes, of weariness and scalding tears and sighs for rest! Why is it? Is this life a stupendous failure? If there is no beyond for which it is preparing, it most certainly is. Could not God have shielded his children from suffering and struggle? Yes: but not without hopelessly excluding them from all prospect of spiritual progress, leaving them forever on the low plane of ignoble, irresponsible brute life. The error is widely prevalent, that God has by some arbitrary decision established the great underlying principles that determine moral character, and can at will change the conditions of spiritual growth. No more mischievous confusion of thought can possibly be entertained. These principles and conditions must reach back infinitely, can of necessity have had no beginning, and cannot be susceptible of the slightest change; for otherwise before their establishment God could not have been possessed of any moral attribute, or have had for his own governance any standard of moral life. He cannot change them or set them aside; for a moment's reflection will disclose that not even he can convert selfishness into a virtue, or place heartless cruelty on a par with a spirit of self-forgetting love.

What he has done for us in this regard is to give power of free choice, and capacity for moral discernment, and to place us in moral relations with himself and with our fellows, and to establish us amid such surroundings as are fitted by their disciplinary processes to develop into glorious fact what are at the first but bare possibilities of virtue. We may, if we choose, stand true to these eternal principles of obligation, live in loving harmony with these many-sided relationships of life, and thereby grow into divine likeness, or we may persistently refuse to conform, and shut against our souls forever this only open door to hope, miss forever this only opportunity to win eternal life. Simply these possibilities are or can be of divine gift. Virtues God cannot bestow : they must be born of battle. Dark as were Christ's forebodings of the coming afflictions of his disciples, deeply as he longed to save them from the imprisonments and scourgings and cruel deaths which awaited them, he, in that last prayer so memorable for its deep, pathetic tenderness, prayed not that his Father would take them out of the world and save them from its sufferings and from its spiritual exposures, but only that he would keep them from the evil, from being finally overmastered and borne down by the terrible power of the tempter. God could not save even his Son, his best beloved. He could by his creative word speak a universe into being, but he could not set aside or render less exacting a single one of the laws of spiritual unfolding, even for Christ himself, though through those long night watches in Gethsemane his shrinking human soul plead for relief with an agony so intense as to cause his body to sweat great drops of blood. Christ, with his human limitations of knowledge, seemed to hope that God might in some way avert the impending doom and still accomplish the objects of his mission, and so he prayed, "Father, if it be possible, let this cup pass from me." Yet while God could not save him from that hour, he no doubt whispered words of comfort, gave assurances of his deep-felt sympathy, promised his loving presence and sustaining grace through it all, and, once his mission ended, a glad and honored welcome to the skies.

What God did for Christ and for his disciples he will do for us, and for this we may most confidently pray, that he will not suffer us to be tempted above that we are able to bear, but will with the temptation provide some way of escape, some way to glorious and final victory. His purpose is to supplement, not supplant. He will send angels to minister, will grant moments of respite, and glimpses of glory.

Our prayers must thus, not only be reasonable, but they must also be offered in the right spirit. The want must be deeply felt, and there must be a whole-soul earnestness in the plea, accompanied with a willingness to make any exertion, and undergo any sacrifice, for the attainment of the end. Until this be our attitude, we are not yet worthy of the help, are not in the mood to appreciate it, and have not the capacity to appropriate its blessings: neither have we prepared the way for God's interference, as we have not fully exhausted our own resources, and thus disclosed the fact, the amount, and the nature of our need. Our prayers should therefore be premeditated, should embody only what we intently long for, what we are convinced we truly require, what after repeated trial we find otherwise beyond our reach, and what in order to obtain we are willing to sacrifice any lower pleasures that stand in their way.

Having thus, after most careful reflection, determined the nature of our requests, being willing to pay the cost involved in the grant, we should come boldly to our Father, and in full faith plead our cause, and then set about life's duties perfectly confident of a favorable answer.

There must be this childlike faith; for Christ's words

of promise were, "Therefore I say unto you, What things soever ye desire when ye pray, believe that ye receive them, and ye shall have them." Christ demanded it of those upon whom he wrought miracles of healing: "Stretch forth thy hand," "Take up thy bed," "Go wash." In the command to make the effort, there was clearly implied the promise to add the strength; but the effort must be made in most trustful confidence before the divine re-enforcement would come. We with good reason rely implicitly upon the trustworthiness of nature's divinely derived physical forces. We are willing to stake, and in fact do stake again and again, our very lives and fortunes on our belief in their promptly answering to our call the very moment certain conditions are fulfilled, and in the surety we feel in their honoring to the letter the terms of their commission. Why not as confidently rely on that more direct divine force for whose help we pray, for it is in as true a sense conditional, with conditions as exact, and it is as prompt and ready to render service the instant those conditions are complied with? Rest assured not until we throw ourselves as unreservedly on the arm of the Almighty as we do on the operations of these lower delegated forces, and this faith is inwrought into the very texture of our lives, can the blessing come.

To have the right spirit when we pray, we must also have our thoughts purged thoroughly from all forms of selfishness. It would seem that so patent a truth requires not even a statement; but this element presents such protean forms, it is so subtle, assumes so many disguises, borrowing the very livery of heaven, that even the elect are many times self-deceived.

Every reasonable prayer offered thus in a right spirit is certain of favorable answer. The blessings bestowed will be either specifically or substantially what we ask; specifically when the objects sought prove to be or to embody what they seem. This is not always, and perhaps not often, the case; and because of that, the blessings are substantially rather than specifically granted. To illustrate: I remember some years since noticing in a show-window what appeared to be a basket of most luscious fruit. The forms and the delicate shadings were remarkable facsimiles of nature's handiwork. The bloom was on the peach and the plum and the purple cluster. On the cheek of the apple glowed those brilliant sunset tints we so admire. The rich, juicy look of the sliced melon was brought out most marvelously. It was a masterpiece of art. I have often thought how differently my little boy, had he been with me, would have looked on this overflowing basket. To him it would have been a complete deception, and he no doubt would have plead with me to make him the happy possessor of it,—not that he might feast his eyes, but his palate. The cool flavors, not the colorings and curves of beauty, would have filled his fancy. A specific answer to his plea would have been a downright disappointment, a disillusion, which he would not at all have relished, for he would have found it but a cunning device of paint and plaster. To have obtained for him the fruit itself, of which he saw only a skilful imitation, would have been to have answered his prayer substantially and to have satisfied his real longings.

Many point to the case of President Garfield as a notable instance of the failure of the prayer test. Countless petitions went up from loving and anxious hearts for his recovery, and yet he died. Because God did not answer these prayers specifically, it is strenuously contended that he did not answer them at all. But how can we, with our extremely limited knowledge, pronounce intelligently on a matter so complicate, involving so many interests, personal, domestic, and national? Is it not possible that God conferred substantially the blessings sought, and that the profits and pleasures which we supposed would flow from Garfield's continuance in the private home circle and in his exalted post of public service were absolutely insignificant compared with what his martyrdom could under divine guidance be made to yield? God very easily could have thwarted the fell purpose of the assassin, and that vast volume of agonizing prayer would never have ascended to his throne from this stricken people. But do you not remember how that event melted into most loving sympathy the hearts, not only of all sections of this great nation, but of all the civilized countries on the globe? Garfield's suffering and death gave to this generation, under God's beneficent overruling, a spiritual impetus and exaltation which this eminent statesman, through a life however long and prosperous, might never have secured. That prayerful and nobly sympathetic attitude of all good people unquestionably made it possible, as nothing else could, for God to thus convert this seeming catastrophe into a most blessed benefaction.

Perhaps he saw such combination of qualities in Garfield's character and in the character of his counselors as to him seemed ominous of evil. There is many a danger signal which we do not detect, or even suspect to exist. It may be, too, God thus sought to impress upon us again one of those lessons taught in President Lincoln's sudden death, just as the terrible war-clouds were lifting, that a nation's strength and safety depend not upon any frail human life, but upon the cherishing of right principles and the continuance of the divine care. For our earthly bereavements and losses we may, if we will, secure priceless compensations, "for our light affliction, which is but for a moment, worketh for us a far more exceeding and eternal weight of glory."

What deep peace has come, and will come still as the years go by, to that once weeping home circle, through the ever sacred memories of the dead! What fondly cherished hopes have been awakened of glad reunions in that golden by-and-by!

The results to President Garfield himself of his weeks of suffering, and final exchange of worlds, while right at the very zenith of his power and his popularity, we have very inadequate means of measuring; for directly behind him, as he answered the summons, there fell an impenetrable veil of mystery. Perhaps, when we too have crossed the river, we shall find that those prayers for life were answered by the gift of larger, grander life than he in his loftiest moods had ever dreamed of getting.

It frequently occurs that most earnest prayers are offered to promote what appear to be directly antagonistic interests. This fact came out very prominently during our late Civil War. For each of the fiercely contending armies, victory was passionately plead for by most devout believers. Who would question the sterling integrity or religious fervor of Stonewall Jackson? and, as we well know, he fought as he prayed. He imperiled his life and finally gave it as a noble sacrifice to the Southern cause. Were his prayers unavailing? Did God turn a deaf ear to the pleadings of this earnest, self-sacrificing disciple? Most assuredly not, though specifically his prayer was denied. Those who fought with him side by side, and shared his local loves and aspirations, but who have been spared to see this day and to enjoy the phenomenal prosperity of the New South,— its quickened pulse, the development of its inexhaustible mineral resources, the birth of its gigantic manufacturing enterprises, its improved agriculture, its rapidly growing cities, its business boom everywhere, and, more than all, its intellectual and moral renascence, and the ushering in of a new era of permanent peace, of genuine fraternal feeling, binding it in indissoluble union with those whom it once faced as foes on stricken fields,—those who have thus lived to see this day, with its rich blessings already realized and with its assured prophecies of vastly multiplied prosperities, recognize now that God, while he swept away their cherished institution of slavery and denied them Southern autonomy, suffered their land to be overrun with devastating war, their homes to be left desolate, and their once proud banners to be torn by cannon shot and trailed in the dust, not only granted them the real blessings which they sought, but multiplied them ten thousandfold. They lamentably erred, as they are now free to confess, as to the channels through which those blessings could come, and they have lived to thank God that he, in his deeper wisdom and in his larger love, himself chose the means through which he should bestow his gifts.

We have discovered in the physical universe multitudes of deadly poisons, hidden under various disguises, bearing remarkably close resemblance to substances that are useful and life-giving. Many of them elude our senses altogether. We fail even with our microscopes and our most careful chemical tests to tear off their masks. We learn of their presence only by their alarming mischief-making. How may of our serious diseases are traceable to these inimical forces, that lurk in the air and water, in the vegetable and animal foods, which we take into our systems unsuspectingly! We are also exposed to intellectual and moral poisons as subtle, as concealed, as deadly, as these which threaten us in the world of matter. How true it is, we are "but children crying in the night, crying for the light, and with no language but a cry," so little certain knowledge have we of what will do us good! and yet, with what unseemly haste we let go our faith, and think our prayers unheard, so soon as any of these hidden poisons are denied!

I remember reading in my early school days, in one of the text-books, of a nobleman, who, while on his return from a long hunt with his favorite hawk on a hot summer's day, filled his cup from a sparkling rivulet that was leaping down the sides of the mountain. As he was lifting it to his parched lips, his hawk with sudden sweep of wings dashed it from his hand, and then, with a strange, anxious call, flew along the bank of the stream toward its source. The nobleman, no little annoyed, again essayed to drink; but the bird the second time upset the cup, and fluttered and called along up the mountain side the same as before. A third time the cup was lifted, and a third time its coveted contents were spilled. The hunter, tired and thirsty, his patience gone, with quick resentment struck his bird a fatal blow. Then, as he looked on his favorite, dead at his feet, it occurred to him to follow up the stream, for the strange conduct of the bird and his strange call had at last impressed him. In the spring, at the very fountain head, he found, to his utter horror, the half-decayed carcass of a huge serpent, and it flashed upon him that it was deadly poison he had been lifting to his lips, that the faithful bird had saved his master's life, and that this same master in a fit of blind passion had ruthlessly destroyed his. Full of remorse, he dug a grave, laid the bird tenderly in it, and afterward, to mark the spot and tell of his gratitude and his grief, he raised a marble shaft above this his humble benefactor. Is there not a lesson here for us? When we are baffled and beaten back in some of our cherished purposes, when the cups of sparkling pleasure which we are eagerly raising to our parched lips are dashed from us, let us not in our haste conclude that our prayers are unblessed, that God has either turned away in deaf indifference and left us to our fate, or become our covert foe. The seemingly hostile forces may be the very angels of his kindest providence, commissioned to smite from our lips by the beating of their strong pinions sparkling drafts which have come from poisoned springs.

With these explanations I reaffirm with added emphasis that every reasonable prayer offered in a right spirit is certain of favorable answer. To this, as we have seen, science can urge no valid objection. It is in consonance with the soundest philosophy; it is in fulfillment of divine promise; it responds to the deepest intuitions of human hearts.

The first effect of modern scientific inquiry has been to weaken faith, and make God seem simply an impersonal, great First Cause, rather than a present loving Father, and ourselves but processes in a vast evolution, parts in an unchangeable order, wheels and pinions, merely, in a mechanism whose movements reach from motes to sun clusters. A reaction from this paralyzing scepticism has already set in. A faith fervent as that felt before science had birth, seems destined again to prevail, and to be the outcome of this very spirit of inquiry which for the past few decades has threatened to relegate it forever to the limbo of the world's outgrown and discarded thought. Reappearing this time as the ripe result of this nineteenth century's tireless and fearless research into time's deepest mysteries, I cannot see how ever again it can lose its hold on the hearts of men.

CHAUTAUQUA
LITERARY AND SCIENTIFIC
CIRCLE

HOME READING COURSE

For 1893-4

Roman History and the Making of Modern Europe
In Politics, Literature, and Art

OFFICERS.

JOHN H. VINCENT, *Chancellor.*   LEWIS MILLER, *President.*
JESSE L. HURLBUT, *Gen'l Sup't.*   KATE F. KIMBALL, *Executive Sec'y.*

COUNSELORS.

LYMAN ABBOTT.   H. W. WARREN.
JAMES M. GIBSON.   W. C. WILKINSON.
EDWARD E. HALE.   J. H. CARLISLE.

A. M. MARTIN, Pittsburgh, Pa., *General Secretary.*
MRS. MARY H. FIELD, San José, Cal., *Secretary for Pacific Coast.*
MRS. A. M. DRENNAN, Ueno, Iga, Japan, *Secretary for Japan.*
MISS M. E. LANDFEAR, Wellington, *Secretary for South Africa.*

THE CHAUTAUQUA CIRCLE.

Aim.

The C. L. S. C. (Chautauqua Literary and Scientific Circle) aims to promote habits of reading and study, in history, literature, science, and art; to give college graduates a review of the college course; to secure for those whose educational privileges have been limited, the college student's general outlook upon the world and life, and to encourage close, connected, persistent thinking.

Plan.

A definite course covering four years.
Each year's course complete in itself.
Specified volumes approved by the counselors.
Allotment of time by the week and month.
A monthly magazine with additional readings and notes.
A membership book with review outlines and other aid.
Individual readers may have all the privileges.
Local circles may be formed by three or more members.
Time required, about one hour daily for nine months.
Certificates granted to all who complete the course.
Seals to be affixed to the certificate are granted for collateral and advanced reading.

Spirit.

The C. L. S. C. maintains that the higher education should be extended to all, young and old, rich and poor, and that education, best begun in academy, college, and university, is not confined to youth, but continues through the whole life. The Circle is not in any sense a college either in its course of study or in its methods of work. Yet it puts into the homes of the people influences and ambitions which will lead many thousand youths to seek colleges and universities. The Circle is unsectarian and unsectional, promoting fraternity and inspiring help to the Home, the Church, and the State.

For whom Designed.

The C. L. S. C. is for busy people who left school years ago, and who desire to pursue some systematic course of instruction.

It is for high school and college graduates, for people who never entered either high school or college, for merchants, mechanics, apprentices, mothers, busy housekeepers, farmer boys, shop girls, and for people of leisure and of wealth.

Many college graduates, ministers, lawyers, physicians, and accomplished women are taking the course. They find the required books entertaining and helpful, affording a pleasant review of studies long ago laid aside. Several members are over eighty years of age; comparatively few are under eighteen. Since 1878, when the Circle was founded, 210,000 readers have joined.

Are you Satisfied with Life?

Is it too late for you to go to school or college (are you too old, or too poor, or too busy)? Should you like to turn mature years, middle life, and old age into youth again? Should you like to turn street, home, shop, railway-car, kitchen, seaside, and forest into recitation rooms? The C. L. S. C. will help you to gratify this desire.

Arrangement of Classes.

The C. L. S. C. was organized in 1878. The class that joined then read four years—that is, 1878–1882. In 1882 this class was graduated, and is still known as the "Class of 1882."

The readings of the several classes for any one year are the same. The course marked out below for the year beginning in the autumn of 1893 and closing in the early summer of 1894 will be: The *first* year for the Class of 1897. The *second* year for the Class of 1896. The *third* year for the Class of 1895. The *fourth* year for the Class of 1894. The class entering in 1893 is the Class of 1897.

Four Years' Course.

1893-94.	1895-96.
Roman and Mediæval History.	American History.
Latin Literature.	American Literature.
Roman and Mediæval Art.	American Government.
Mediæval Literature.	Social Institutions.
Political Economy.	Physiology.
Religious Literature.	Religious Literature.

1894-95.	1896-97.
English History.	Greek History.
English Literature.	Greek Literature.
English Composition.	Greek Art.
Astronomy.	Ancient Greek Life.
Geology.	American Diplomacy.
Religious Literature.	Religious Literature.

Required Literature.

The Circle has gradually secured a class of books written by leading authors, and especially adapted to the needs of self-educating readers. *The Chautauquan*, organ of the C. L. S. C., contains much of the required reading for each year, and many timely articles by the best American and English writers.

Prescribed Reading for 1893-94.

ROME AND THE MAKING OF MODERN EUROPE, James R. Joy......$1.00
ROMAN AND MEDIEVAL ART, William H. Goodyear.......................... 1.00
OUTLINES OF ECONOMICS, Richard T. Ely.. 1.00
CLASSIC LATIN COURSE IN ENGLISH, W. C. Wilkinson............... 1.00
SONG AND LEGEND FROM THE MIDDLE AGES, Edited by W. D. McClintock .... .50
SCIENCE AND PRAYER, Rev. W. W. Kinsley.................................... .50
THE CHAUTAUQUAN (12 numbers).. 2.00

The Chautauquan Magazine

Will contain illustrated articles on European Life in the Middle Ages, American Colonies in the Continental Capitals, the influence of Roman language, literature, and art on our own times, and papers on a wide range of present day topics.

Memoranda.

The membership book contains duplicate sets of question papers, called memoranda. These are not examination papers but are review questions which may or may not be answered from memory. *The filling out of these memoranda is not essential to graduation.*

The four-page paper gives a brief condensed review of the whole course, and members who fill out this paper for each of the four years, receive one white seal at graduation. The twelve-page paper offers a more thorough review. One white seal is given for *each paper* which shows 80 per cent of correct answers. Besides the seal,

(1) Any seal course paper will be *corrected and returned to the student* upon payment of a special fee of 50 cents.

(2) Any seal course paper will be *graded and returned to the student* upon payment of a fee of 25 cents. (In this case the questions which are not wholly correct will be indicated but the correct answers will not be given.)

(3) All other *seal* papers for which no special fee has been paid will be graded and the exact *grade* reported to the student but the papers will not be returned. The four-page papers will be examined to determine whether they rank above or below eighty and the result reported.

(4) The four-page papers will be graded and returned for a fee of twenty-five cents, or corrected and returned for a fee of fifty cents. One fee for the four papers.

How to Join the Circle.

Send answers to the following questions together with *fifty cents* (fee for one year) to *John H. Vincent, Drawer 194, Buffalo, N. Y.* [A blank containing these questions may be had by applying to the Buffalo Office.]

1. Give your name in full. 2. Your post-office address, with county and state. 3. Are you married or single? 4. What is your age? Are you between twenty and thirty, or thirty and forty, or forty and fifty, or fifty and sixty, etc.? 5. If married, how many children living under the age of sixteen years? 6. What is your occupation? 7. With what religious denomination are you connected? 8. Are you a graduate of a High School or College? If so, give the name of the institution. 9. If you have been a member of the C. L. S. C. in past years, but are now beginning anew, state to what Class you formerly belonged. 10. Do you join as (a) an individual reader, (b) a Home Circle reader (in a family), or (c) as a "Local Circle" reader? The reader may change from one relation to another at will.

The Class of 1897 will be organized during the autumn of 1893, but students will be received at any time.

How to Obtain the Literature.

All the required literature (books and THE CHAUTAUQUAN) may be obtained by sending a draft or money order for $7 to Flood & Vincent, The Chautauqua-Century Press, Meadville, Pa. On all orders of five or more sets of books sent to the same address by express (charges unpaid) a discount of ten per cent will be allowed. Books singly and THE CHAUTAUQUAN separately if desired.* (To foreign subscribers in countries included in the postal union, THE CHAUTAUQUAN will be sent, postpaid, for $2.60, to South Africa, for $3.24.)

Membership Fee.

1. To defray expenses of correspondence, membership book, etc., an annual fee of fifty cents is required. This amount should be forwarded to *John H. Vincent, Drawer 194, Buffalo, N. Y.*, by New York draft, post-office order, or postal note.

2. In sending your fee be sure to state to which class you belong, whether 1894, 1895, 1896, or 1897. A special blank is furnished to secretaries of local circles who forward fees.

---

* Subscriptions for THE CHAUTAUQUAN alone should be addressed to Dr. T. L. Flood, Meadville, Pa.

3. Keep a record of every order sent, including date, names, and amount.

4. Before forwarding a post-office order or postal note, see that it is properly dated, drawn for the right amount, and made payable at Buffalo, N. Y. In regard to diploma fee, see C. L. S. C. Hand-Book, § 9.

Fee for Graduates.

The following simple **arrangement** has been made for graduates (Classes of '82–93) who wish to pursue the current year's course of reading—with the undergraduates:

An annual fee of 50 cents will entitle a graduate to all communications from the Central Office for that year, including the twelve-page memoranda on the regular year's reading.

In this way two seals can be earned:

1. For reading the books of the *regular* course and filling out the regular four-page memoranda, a *special seal*.

2. For filling out the twelve-page memoranda on the reading of the regular course, a white seal will be given, if 80 per cent of the questions are correctly answered. See also sections (1), (2), and (3) of "Memoranda" paragraph, page 4.

Recommended Order of Study for 1893-94.

(For Readers beginning October 1, 1893.)

October.
Rome and the Making of Modern Europe—Joy—to page 62.
Outlines of Economics.
THE CHAUTAUQUAN.

November.
Rome and the Making of Modern Europe—to page 117.
Economics.
THE CHAUTAUQUAN.

December.
Rome and the Making of Modern Europe—to page 174.
Economics.
THE CHAUTAUQUAN.

January.
Rome and the Making of Modern Europe—to page 260.
Economics—finished.
Roman and Medieval Art—Goodyear—to page 111.
THE CHAUTAUQUAN.

February.
Rome and the Making of Modern Europe—concluded.
Roman and Medieval Art—to page 194.
THE CHAUTAUQUAN.

March.
Classic Latin Course in English—Wilkinson—to page 90.
Roman and Medieval Art—concluded.
Song and Legend from the Middle Ages—McClintock—to page 37.
THE CHAUTAUQUAN.

April.
Classic Latin Course in English—to page 244.
Song and Legend—to page 112.
THE CHAUTAUQUAN.

May.
Classic Latin Course—concluded.
Song and Legend—concluded.
Science and Prayer - begun.
THE CHAUTAUQUAN.

June.
Science and Prayer.
THE CHAUTAUQUAN.

Local Circles

Individuals may prosecute the studies of the C. L. S. C. alone, but their efforts will be greatly facilitated by securing a local circle of two or more persons who agree to meet as frequently as possible, read together, converse on the subjects of study, arrange for lectures, organize a library, a museum, a laboratory, etc. A local circle may give attention to the cultivation of taste, cleanliness, etc., in towns and villages, and discuss sanitary and other questions tending to public health and social progress.

All local circles should, as soon as organized, report the names of their officers to *John H. Vincent, Drawer 194, Buffalo, N. Y.* Several pages of THE CHAUTAUQUAN are devoted especially to the interests of the circles, but none are recognized in the magazine unless they report to the Buffalo Office.

Many circles include in their membership local members—students who, not having paid the membership fee, are not enrolled at the Central Office, but who, nevertheless, read much of the prescribed course and attend the meetings of the Circle. It is hoped that all interested in the C. L. S. C. will become, if possible, regular members, that while enjoying its benefits they may also contribute to its support.

Organizers of Circles.

The Central Office solicits correspondence with school teachers, ministers, and others who wish to promote an interest in intellectual work in their communities. Intelligent, enthusiastic leaders are essential to successful local work, and their coöperation is earnestly desired.

C. L. S. C. Mottoes.

"*We study the Word and the Works of God.*"
"*Let us keep our Heavenly Father in the midst.*"
"*Never be discouraged.*"

What the C. L. S. C. has Done for Some of Its Members.

From a School Teacher.

"Last year I read alone, but this year I have succeeded in getting three others to join me. We meet weekly, read aloud a portion of the lesson, have informal talks and enjoy it thoroughly. We are deriving great benefit as well as establishing a custom of forming reading circles among all classes of persons. I am a busy school teacher, and the reading is as a tonic. It lifts me out of my old life. It fills me with inspiration and determination to 'Look Up and Lift Up.' I cannot say enough in praise of the C. L. S. C."
——, Kansas.

From an Isolated Reader.

"We have just finished our third course in the Chautauqua Circle—my two sisters and myself. Allow us to express our heartfelt gratitude for the great blessing this Circle has brought to our home. It is indeed a great boon to us. We live in a remote neighborhood where there are no schools and have been deprived of the advantages of a college education, and are great lovers of literature. From this you may have some idea of the blessing the Chautauqua Circle is to us."——, Alabama.

From Mothers.

"I am the mother of eight children, and have done my own work during the four years with the usual amount of sickness that follows such a family. My cares have been great, yet I would not be the woman I am to-day had it not been for the C. L. S. C. work that has employed my mind in thinking of better things than the everyday cares of life. I hope other tired mothers will be benefited as I have been."

"I am the wife of a farmer who works 300 acres; keep a girl only six months in the year. I have a great deal to see to and to do. I wanted to take up the C. L. S. C. course two years before I did, but thought I would wait until I could have more time. I gave up waiting for time and just took it."

www.ingramcontent.com/pod-product-compliance
Lightning Source LLC
Chambersburg PA
CBHW020129170426
43199CB00010B/694